CAMBRIDGE PHYSICAL SERIES

EXPERIMENTAL PHYSICS

A TEXT-BOOK OF
MECHANICS, HEAT, SOUND AND LIGHT

EXPERIMENTAL PHYSICS

A TEXT-BOOK OF
MECHANICS, HEAT, SOUND AND LIGHT

BY

HAROLD A. WILSON, M.A., D.Sc., F.R.S.,

formerly Fellow of Trinity College, Cambridge, England;
Professor of Physics in the Rice Institute, Houston, Texas, U.S.A.

Cambridge :
at the University Press
1915

CAMBRIDGE
UNIVERSITY PRESS

University Printing House, Cambridge CB2 8BS, United Kingdom

Published in the United States of America by Cambridge University Press, New York

Cambridge University Press is part of the University of Cambridge.

It furthers the University's mission by disseminating knowledge in the pursuit of education, learning and research at the highest international levels of excellence.

www.cambridge.org
Information on this title: www.cambridge.org/9781107637672

First published 1915
First paperback edition 2014

A catalogue record for this publication is available from the British Library

ISBN 978-1-107-63767-2 Paperback

PREFACE

THIS book is intended as a text-book for use in connection with a course of experimental lectures on mechanics, properties of matter, heat, sound and light. No previous knowledge of physics is assumed, but nevertheless the book is primarily intended for a first year college course, and the majority of the students attending such a course have studied elementary physics at school. The writing of such a book does not offer much scope for originality; the aim of the writer should be to present fundamental principles clearly and accurately. The chief difficulty is to decide what to include and what to leave out. I have endeavoured to leave out everything not of fundamental importance. It is important for the student to learn some facts and to get to understand some methods and fundamental principles; if he learns nothing about certain phenomena no harm is done and he can make up the deficiency in his knowledge at a later date if necessary. The kind of text-book which contains a little about everything does more harm than good.

Care has been taken not to discuss questions which cannot be treated adequately in an elementary way and to avoid stating formulae without proving them. A few experiments are rather fully described in nearly every chapter; these have been selected from the many which might have been merely mentioned.

In Part I, Chapters VI, VII and parts of IX may be omitted at the first reading. In Part II, Chapters X and XI may also be omitted by students whose time is limited.

I am indebted to Messrs J. J. Griffin and Sons, Ltd. for permission to reproduce Fig. 57, Pt I (Oertling balance), from their catalogue; to the Cavendish Laboratory for leave to draw Fig. 62, Pt I; to Messrs G. Cussons, Ltd, for Fig. 82, Pt I; to Messrs W. G. Pye and Co. for Fig. 22, Pt II, and to Mr Edward Arnold for permission to reproduce Fig. 54, Pt IV from Schuster's *Optics*.

I wish to express my thanks to Mr T. G. Bedford, the Editor of the Cambridge Physical Series, for many valuable suggestions and corrections in reading the proofs and for the preparation of Fig. 24, Pt III.

My thanks are also due to the staff of the University Press for their excellent work.

H. A. W.

July, 1915.

CONTENTS

PART III

SOUND

PART IV

LIGHT

INTRODUCTION

THE physical sciences relate to the study of phenomen
systems of bodies composed of lifeless or inanimate matter.
name *physics* denotes a restricted branch of physical scie
dealing usually with only the simplest kinds of phenomena, whi
happen under artificially arranged circumstances. These circun
stances are designed with the object of simplifying as far a
possible the actions taking place, so that the real nature of the
phenomena observed may be the more easily discovered. Physics,
then, may be said to relate to the succession of events in comparatively simple material systems artificially contrived for special
purposes.

The other great branch of physical science, chemistry, deals
chiefly with the composition and special properties of particular
substances and their methods of preparation in a state of purity.

All man's knowledge of things is the result of experience. To
study physical science is to acquire for one's own use the ac-
cumulated experience and wisdom of mankind in dealing with
nature and trying to turn the manifold properties of matter to
the good of the race.

Experience shows that natural phenomena are subject to
definite rules or laws which are invariable. Events do not occur
in a haphazard manner, but follow each other in regular order.
Any given event is determined by the state of things preceding it
according to definite laws. If a given state of things at any time
or place is followed by a certain event, then at any other time and
place a precisely similar state of things will be followed by an
identical event. On this foundation science rests, and without it
no reliable knowledge would be possible. It is sometimes said
that like causes produce like effects, and this is true if *like causes*
means causes which differ only with respect to the time and place.
It is also true that in many cases a small change in a system
produces only a small change in the events taking place in it;

but there are other cases in which this is not true, for example a small movement of a gun just before it is fired.

Clerk-Maxwell stated the general maxim of physical science in the following words*:—

"The difference between one event and another does not depend on the mere difference of the times or the places at which they occur, but only on differences in the nature, configuration, or motion of the bodies concerned."

Our belief in the truth of this maxim is based on experience. No exceptions to it, supported by reliable evidence, are known.

To describe phenomena exactly it is necessary to use words with definite meanings. Many words are used in ordinary conversation in a loose manner sometimes with one meaning and sometimes with another. In scientific work an attempt is made to formulate precise definitions of the meanings of words, and when a word has been given a meaning it should never be used with any other. One of the chief advantages of a scientific training ought to be the ability to use language having only one possible meaning.

To measure a quantity it is necessary to have a unit in terms of which the size of the quantity can be expressed. The size of any quantity is a unit of the same kind as the quantity itself multiplied by a number. The choice of suitable units of exactly fixed magnitudes is one of the most important aids to scientific progress.

The discovery of new phenomena and the exact measurement of all the quantities concerned is one of the chief objects of scientific investigations. It is also desired to find out the general laws regulating phenomena, to analyse complicated phenomena into simpler components, and to formulate mental pictures of the innermost structure of things, so that the succession of events may be explained and the behaviour of matter in given circumstances predicted.

When matter is found under some circumstances to act in accordance with a certain rule or law, this rule may be suggested as a universally true law of nature. The consequences to be expected if the rule is always obeyed are then worked out as completely as possible, and compared with the results of observation

* *Matter and Motion*, p. 21.

When the results in an immense number of cases have been found to agree with those predicted, and when no cases of want of agreement are known, the rule comes to be regarded as an established law of nature.

When an event is shown to have occurred in accordance with previously established laws of nature, so that it could have been completely predicted by assuming the laws to be obeyed, then it is sometimes said to have been explained in terms of the natural laws in question. Such so-called explanations are of course not complete. The event has merely been given its proper place in a class of similar events determined by the same laws. A complete explanation would require the laws themselves to be explained. No such thing as a really complete explanation of any event can be given.

If it could be shown that all phenomena were due to the motion of a single continuous medium filling all space and that nothing else existed in the universe, the laws of motion of the medium would still remain to be explained.

Many of the properties of matter depend on the structure of parts far too small to be observed directly. To explain such properties a mental picture of the small parts may be imagined, and they may be supposed to obey certain laws. The truth of such an hypothesis or theory can be tested by comparing the properties of matter in bulk with the properties to be expected according to the theory in question. Even if the expected properties agree perfectly with those observed, we can never be sure that the assumptions made are really true in fact. Other assumptions might lead to identical properties.

Such theories of the nature of matter enable new phenomena to be predicted, and when such predictions are found true the truth of the theory becomes more probable. When new phenomena are discovered an attempt is generally immediately made to formulate a more or less complete theory to explain them. This theory then serves as a guide in devising new experiments with the object of elucidating the real nature of the phenomena. Thus the theory serves a useful purpose even if it ultimately proves to be wrong.

When a theory enables a great body of facts to be explained, when it has successfully predicted new facts, and when no facts inconsistent with it are known, it is usually regarded as probably more or less completely true. For example the atomic theory, according to which each elementary substance is made up of an immense number of minute parts all exactly equal in size and structure, explains successfully such a vast array of well-established facts that it is universally accepted as true.

When two different theories are capable of explaining a set of facts, then to decide between them their consequences are worked out until some practical case is found for which they predict different results. The matter is then tested by experiment, and if the consequences predicted by one of the theories are in agreement with the results obtained, the other theory must be abandoned or modified. Such an experiment to decide between rival theories is called a *crucial experiment*.

In studying physics it is important to distinguish between simple but inexact experiments designed merely to illustrate well-established principles and the exact investigations by which such principles have been established. For example, though Atwood's machine enables the laws of motion to be illustrated, our belief in their universal truth does not rest on such a crude basis but is founded on the results of many series of investigations of high precision.

It is convenient to divide physics into several parts, each dealing with closely related phenomena.

Mechanics is the study of the motion of matter.

Under the heading *General Properties of Matter* are included gravitation, elasticity, capillarity, viscosity, and other miscellaneous phenomena.

Sound is the part of physics which deals with the vibrations of elastic bodies which produce sensations in our ears.

Heat is the branch of physics dealing with all phenomena depending on whether bodies are hot or cold.

Light relates to the physical process which produces the sensation of sight and is closely related to another great division of physics, *Electricity* and *Magnetism*.

REFERENCES

Matter and Motion, J. Clerk-Maxwell.
The Principles of Science, W. S. Jevons.

PART I

MECHANICS AND PROPERTIES OF MATTER

CHAPTER I

SPACE AND TIME

MECHANICS is the study of the modes of motion of matter. The motion of matter involves the ideas of position, time, and quantity of matter or mass.

The position of a body can be described by stating the length of the straight line between it and another body whose position is known and the direction of this line. Thus if we are told that a village is 20 miles from the town where we live and due West of it, we know the position of the village. The position of a body can only be described relatively to the positions of other bodies. The position of a single particle by itself in space could not be specified, because there would be nothing from which its distance could be measured; for all parts of space are exactly similar. With two particles in space the distance between them could be measured and would give the position of either relative to the other. When a person gets to know the relative positions of the chief objects in his neighbourhood, such as the buildings and streets in the town where he lives and the different things in his house, he has acquired by experience the ideas of position and of space or volume. The idea of space is acquired by experience; it cannot be explained. Any particular body occupies a certain portion of space.

The idea of space carries with it the ideas of length, breadth, and height, or more generally, of the possibility of moving in three directions perpendicular to each other. Only three straight lines can be drawn through a point so that each one is perpendicular to the other two. On this account space is said to have three dimensions.

Distances are measured in terms of a unit of length. The standard yard is a metal bar, having two narrow transverse lines ruled on it, which is carefully preserved in London, England. The distance between the centres of these two lines when the bar is surrounded by melting ice is the English unit of length, the yard. Copies of this standard, very nearly equal to it, are made, and are used as secondary standards with which ordinary yard measures for general use may be compared. The Bureau of Standards at Washington, U.S.A., possesses such copies, whose lengths have been accurately compared with the original standard.

The French standard of length, called the metre, is preserved at Sèvres, and consists of a similar bar. One thirty-sixth part of a yard is one inch, and the metre is equal to 39·370 inches. One inch is equal to 2·5400 centimetres.

To find the length of any body it is necessary to determine how many times a unit of length is contained in its length.

If L denotes a unit of length, such as one yard, one metre, or one centimetre, then any length is equal to a number n times the unit or nL. For example consider a length of 100 cms. or 39·37 inches.

Units of area and volume are derived from the units of length. The unit of area is the area of a square with sides of unit length. The unit of volume is the volume of a cube with sides of unit length. If L denotes a unit of length then the corresponding units of area and volume may be denoted by L^2 and L^3 respectively, for the unit of area varies as the square of the unit of length and the unit of volume as the cube.

The area of a rectangle, with sides of lengths nL and mL is equal to $(nL) \times (mL)$ which may be written nmL^2 and denotes nm units of area. The arithmetical operation indicated by the expression nmL^2 is the multiplication of the number n by the number m. If $n = m = 1$, the expression nmL^2 reduces to L^2 and denotes one unit of area; the arithmetical operation indicated by L^2 being the multiplication of the number one by the number one. In the same way the arithmetical operation indicated by L^3 is $1 \times 1 \times 1$.

The idea of time is acquired by our experience of the succession of events, especially of such events as the rising and setting of the sun, which are repeated in a regular manner. The rotation of the earth about its axis, so far as we know, is not retarded by any appreciable forces and is therefore supposed to go on at a practically constant rate. In consequence of this uniform rotation the stars appear to revolve round the earth. The time of one complete revolution of a star round the earth is called a sidereal day. The sun also appears to revolve round the earth, but owing to the motion of the earth relative to the sun, the time of one revolution is not exactly constant. The time of one revolution of the sun round the earth is called a solar day. The variation in the lengths of the solar days can be measured by comparing them with the sidereal days, which are taken to be all equal. The average length of a solar day is called a mean solar day, and is the fundamental unit of time adopted for all practical purposes. The second is $\frac{1}{86400}$ of a mean solar day. Time is indicated by clocks, which should be regulated so that they indicate 86400 seconds in a mean solar day. The standard clocks in astronomical observatories are checked by observations on the apparent motions of the sun and stars. If T denotes a unit of time such as one second, then any time interval is equal to nT, where n is a number.

The mass of a body may be said to be the amount of matter in it. The English unit of mass is the mass of a certain piece of platinum carefully preserved in London and called the standard pound. Copies of this standard, of nearly equal mass, can be obtained. The French unit of mass is the mass of a piece of platinum, kept at Sèvres, called the standard kilogram. The gram is one-thousandth part of a kilogram. One pound is equal to 453·6 grams. If M denotes a unit of mass such as a gram, then any mass is equal to nM, where n is a number.

The magnitude of any physical quantity is equal to a unit,

Fundamental and derived Units.

of the same kind as the quantity, multiplied by a number. If the size of the unit is changed the number varies inversely as the unit. For example, a length of nine feet is equal to three yards because one yard is equal to three feet.

The units of length, time and mass which we have denoted by L, T and M are usually regarded as the fundamental units. Whenever possible other quantities are expressed in terms of units derived from these fundamental units. Thus the unit of volume, which may be denoted by L^3, is derived from the unit of length. As another example consider density. The density of a body is its mass per unit volume. The unit of density is taken to be a density of unit mass per unit volume. A body of mass nM and volume mL^3 has a density equal to nM/mL^3 which may be written $\frac{n}{m} ML^{-3}$ and denotes n/m units of density. The arithmetical operation indicated by nM/mL^3 is the division of the number n by the number m. If $n = m = 1$, the expression nM/mL^3 reduces to M/L^3 and denotes one unit of density. The arithmetical operation indicated by M/L^3 is the division of the number one by the number one or one unit of mass divided by one unit of volume gives one unit of density. Any density is equal to pML^{-3} where p denotes a number. The general expression for any unit, derived from the fundamental units, can be written $L^x M^y T^z$, and of course is numerically equal to one. It is not always possible to derive units from the fundamental units; for example, the unit of temperature is not so derived.

Suppose we have an equation expressing a relation between physical quantities, say $p = q$. p must be a quantity of the same nature as q for it is impossible, for example, to have a volume equal to an area or a density. Consider the equation $m = vd$, where m denotes the mass of a body, d its density and v its volume. m means m times the unit of mass M. v means v times the unit of volume L^3. d means d times the unit of density ML^{-3}. Thus the equation may be written

$$mM = (vL^3)(dML^{-3}),$$

where m, v and d now denote mere numbers. Since $m = vd$ this reduces to $M = L^3 ML^{-3}$ and so to $M = M$ which shows that both sides of the equation represent a mass. The equation might be written

$$m \text{ grams} = (v \text{ cubic cms.}) (d \text{ grams per c.c.})$$

if the fundamental units adopted were the gram and centimetre.

If both sides of an equation cannot be shown to represent quantities of the same kind, in this way, then the equation must be false.

When a given unit varies as the nth power of one of the fundamental units, it is said to be of n dimensions Dimensions. as regards that unit. Thus the expression ML^{-3} for the unit of density indicates that the dimensions of the unit of density, in terms of the fundamental units, are one as regards mass and -3 as regards length.

REFERENCE

Matter and Motion, J. Clerk-Maxwell.

CHAPTER II

MOTION

WHEN the position of a body with respect to surrounding bodies remains unchanged it is said to be at rest. When its position continually changes it is said to be in motion. If a body moves from a position A to another position B, then the straight line AB is called the displacement of the body. A displacement has magnitude and direction. Quantities like volumes and masses which have magnitude but not direction are called *scalars*, while quantities which have magnitude and also direction are called

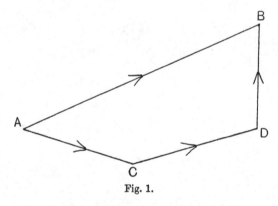

Fig. 1.

vectors. Any vector can be represented by a straight line such that its length represents the size of the vector and its direction the direction of the vector.

If a body is displaced from A to B and from B to a third point C, then AC, the resultant displacement, is called the vector sum, or simply the sum, of the two displacements AB and BC. Any displacement AB (Fig. 1) can be resolved into two or more component parts such as AC, CD and DB; for the resultant of displacements from A to C, C to D and D to B is the same as a displacement from A to B.

If a body moves along at a uniform rate, so that it covers equal distances in equal times, it is said to be moving with a uniform velocity. Its velocity is measured by the distance it describes in unit time. In a time t it will describe a distance s given by the equation $s = tv$, where v denotes its velocity. The unit of velocity is the velocity of a body which travels unit length in unit time and is denoted by $L \div T$ or LT^{-1}. Velocity is a vector. A uniform velocity can be represented by a straight line containing as many

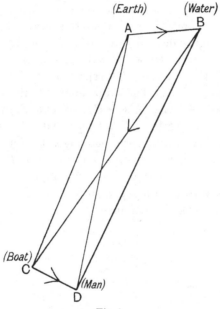

Fig. 2.

units of length as the velocity contains units of velocity and drawn in the same direction as the velocity. Velocity like position and displacement is relative to surrounding bodies. If a body were alone in otherwise empty space it would not be possible to tell whether it was at rest or in motion.

When the objects near a body are themselves moving relative to each other, the velocity of the body relative to one object will not be the same as its velocity relative to another. For example, consider the velocity of a man on the deck of a steamboat. Suppose the man walks along the deck at 4 miles an hour relative to the boat. Let the boat be moving through the water at 20 miles an hour, and let the water be flowing over the earth at, say, 6 miles an hour. If all three velocities are in the same direction, say from North to South, it is easy to see that the velocity of the man relative to the water is $20 + 4 = 24$ miles an hour and relative to the earth $20 + 4 + 6 = 30$ miles an hour. If they are not in the same direction, the velocity of the man relative to the earth can be found in the same way as the resultant of two or more displacements, for the velocities are equal to the displacements in unit time. To do this draw AB (Fig. 2) to represent in magnitude and direction the velocity of the water relative to the earth. Draw BC representing the velocity of the boat relative to the water and CD representing the velocity of the man relative to the boat. Then AD represents the velocity of the man relative to the earth. Also AC represents the velocity of the boat relative to the earth and BD represents the velocity of the man relative to the water. Corresponding to the four bodies, earth, water, boat and man, we have the four points A, B, C and D, and a line from any one point to any other represents the relative velocity of the corresponding pair of bodies. Such a diagram as this is called a velocity diagram.

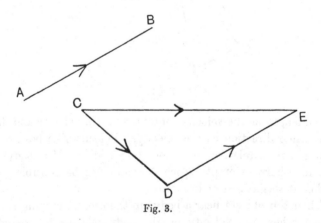

Fig. 3.

All other kinds of vector quantities besides displacements and velocities can be represented by properly drawn straight lines, and can be combined together in the same way. Thus if AB and CD (Fig. 3) represent two vectors of the same kind, their resultant can be got by drawing from D a line DE equal and parallel to AB. Then CE represents in magnitude and direction the resultant or vector sum of CD and DE or of CD and AB. If AB (Fig. 4) represents any vector, then AC and CB represent component parts of the vector AB. If the angle at C is a right angle, then AC is

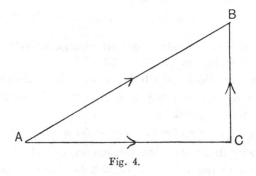

Fig. 4.

called the resolved part of AB in the direction of AC. Let the angle $BAC = \theta$; then $AC = AB \cos \theta$. To get the resolved part of any vector along a direction making an angle θ with it we multiply by $\cos \theta$.

The vector represented by AB is equal and opposite to the vector represented by BA, so that $AB = -BA$.

When a body moves with a changing velocity it is said to have acceleration. In such cases the space described by the body in any interval of time divided by the time gives the average value of the velocity during this time. The equation $s = tv$ is still true if v now denotes the average velocity during the time t. If the time interval t is taken small enough the changes in the velocity during it will become negligible, and then the average velocity will be equal to the actual velocity during the extremely short interval. The velocity at any instant is the space described during an extremely short interval of time containing the instant divided by the length of the interval. The interval taken must be so short that no appreciable change in the velocity takes place during it. It may be one-thousandth or one

Acceleration.

ten-millionth of a second or as much smaller as we please to
imagine it.

When the velocity of a body moving along a straight line
changes at a constant rate the body is said to have a uniform
acceleration. The acceleration is then equal to the change of
velocity in unit time. If the velocity at the beginning of an
interval of time t is equal to v_1, and at the end equal to v_2, then
the uniform acceleration a is given by the equation

$$a = \frac{v_2 - v_1}{t} \text{ or } v_2 = v_1 + at.$$

The unit of acceleration is unit change of velocity in unit
time, it is therefore denoted by LT^{-2}.

The average velocity of a body moving with a uniform accelera-
tion along a straight line is $\frac{1}{2}(v_2 + v_1)$ so that the space described
in the interval t is given by

$$s = \tfrac{1}{2}(v_2 + v_1)t = \tfrac{1}{2}(v_1 + at + v_1)t = v_1 t + \tfrac{1}{2}at^2.$$

If the velocity diminishes, the acceleration must be taken to be of
opposite sign to the velocity. In this case the velocity becomes
zero after a time t' given by

$$0 = v_1 + at' \text{ or } t' = -\frac{v_1}{a}.$$

Thus if $v_1 = +200$ cms. per sec. and $a = -10 \frac{\text{cms.}}{\text{sec.}^2}$ we get

$$t' = \frac{-200}{-10} = +20 \text{ secs.}$$

We have
$$v_2 - v_1 = at,$$
$$v_2 + v_1 = \frac{2s}{t}.$$

Multiplying the corresponding sides of these equations together
we get $v_2{}^2 - v_1{}^2 = 2as$. This shows that the change in the square
of the velocity is proportional to the space described.

If the body starts from rest we have $v_1 = 0$, so that

$$v_2 = at,$$
$$s = \tfrac{1}{2}at^2,$$
and
$$v_2{}^2 = 2as.$$

An acceleration, like a displacement or a velocity, is a vector and
so can be represented by a properly drawn straight line; and the

resultant or vector sum of two or more accelerations can be found by the same rule as the sum of other vectors.

When the direction of motion of a body is changing, the body has an acceleration the direction of which does not coincide with the direction of motion. Suppose at the beginning of a short interval of time t the velocity of the body is represented by AB, and at the end by AC (Fig. 5). Then to change AB into AC

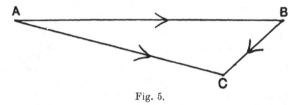

Fig. 5.

it is necessary to add to AB the velocity represented by BC. The average acceleration during the interval is therefore BC/t and its direction is from B to C.

An important case is that of a body moving round a circle with uniform velocity.

At the beginning of an interval of time t let the body be at P
Uniform Motion   (Fig. 6) and at the end suppose it to have reached
in a Circle. Q. Then if v is its velocity, $PQ = vt$. Let PA and QB represent its velocity in magnitude and direction at P and Q respectively. Draw a line PC equal and parallel to QB. AC represents the velocity which added to the velocity at P converts it into that at Q. If we take the time t very short the acceleration will not vary appreciably, so that it will be equal to the change in velocity divided by t. If t is very small the angle POQ, which is equal to APC, will be very small. Let $OP = OQ = r$. We have

$$P\hat{O}Q = \frac{vt}{r}.$$

Hence $AC = PA \times \frac{vt}{r} = \frac{v^2 t}{r}.$

Thus the acceleration is equal to $\dfrac{AC}{t} = \dfrac{v^2}{r}$ and is parallel to AC, that is, it is in the direction from P to O.

It appears therefore that a body moving with uniform velocity v round a circle of radius r has an acceleration equal to v^2/r in the

direction from the body to the centre of the circle. The acceleration is perpendicular to the velocity; and therefore it does not alter
the magnitude of the velocity, but it changes its direction.

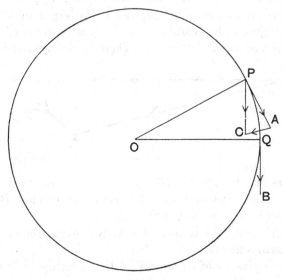

Fig. 6.

Let a point P (Fig. 7) be moving with uniform velocity v
round a circle $BRAQ$. Drop a perpendicular PN

Simple Harmonic Motion. on the diameter AOB of the circle. As P goes
round suppose that N moves along BA so as to
keep always on a line through P perpendicular to BA. Then as
P moves along BRA, N moves along BOA, and as P goes along
AQB, N goes back along AOB. Thus as P goes round and round
the circle, N moves backwards and forwards along the diameter
between A and B. The motion of N is what is called a *simple
harmonic motion*. While P goes once around, N is said to make
one complete vibration or oscillation.

The greatest distance N goes from O, which is equal to the
radius r of the circle, is called the amplitude of the simple harmonic
motion. The number of complete vibrations performed in unit
time is called the frequency of the vibration. Let the time of
one complete vibration be T, and let the frequency be n, so that

$n = 1/T$. We have $2\pi r = Tv$ or $2\pi rn = v$. The acceleration of P is equal to v^2/r and is directed from P to O. We may take PO to represent this acceleration in magnitude as well as in direction. This acceleration represented by PO can be regarded as the resultant of two components represented by PN and NO. Of these PN is perpendicular to AB and so does not affect the motion of N. The other represented by NO is equal to the acceleration of N. Now PO represents an acceleration equal to

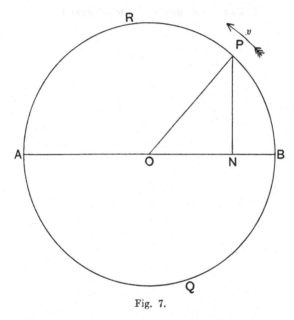

Fig. 7.

v^2/r, so that NO represents one equal to $\dfrac{v^2}{r} \times \dfrac{NO}{PO} = \dfrac{v^2}{r^2} \times NO$. Hence we find that N has an acceleration directed towards O and equal to its distance from O multiplied by v^2/r^2. Let $v^2/r^2 = \mu$. The acceleration of N is then equal to μNO or to $-\mu ON$. We have

$$T = 2\pi \frac{r}{v} = 2\pi/\sqrt{\mu}.$$

Thus it appears that when a point moves along a straight line in such a way that its acceleration is equal to μ times its distance from a fixed point in the line and is directed towards that point, then it oscillates backwards and forwards, and the time of one complete oscillation is equal to $2\pi/\sqrt{\mu}$. If μ is a constant, then the number of vibrations per second is a constant independent of the amplitude of the oscillations. At O the middle point of the swings the velocity of N is equal to that of P, and so is equal to $2\pi r n$.

REFERENCE

Matter and Motion, J. Clerk-Maxwell

CHAPTER III

THE LAWS OF MOTION OF MATTER

IT is found that the way a body moves is modified by the presence of other bodies. There is said to be an action between the bodies which changes their modes of motion. Such actions take place between bodies which are in contact and also between bodies separated from each other. ' For example, the cars on a railway are set in motion by the engine, and the motion of the engine is modified by the presence of the cars. Again, bodies when unsupported fall towards the earth; there is an action between the earth and the bodies near it. A magnet and a piece of iron move together when near each other. The motion of a falling body is changed when it meets the ground; and so on. In all such cases there is a mutual action between two or more bodies which modifies their motion.

An action which causes bodies to come together is called an attraction, while one which tends to separate them is called a repulsion. Thus we say that there is an attraction between the earth and other bodies. We may consider the effect of such actions on one alone of the bodies concerned. The change in the mode of motion of the body considered is said to be due to a force acting on the body, which force is said to be produced by the other body or bodies. Force then may be defined as that which changes the motion of a body. The mutual action between two bodies changes the motion of both, so that each exerts a force on the other. The two forces are merely two different aspects of the one mutual action.

If a horse pulls a boat along by means of a rope, there is an action between the horse and the boat which is transmitted by the rope. The rope is said to be in a state of tension. Consider the two parts of the rope on either side of any cross section P (Fig. 8)

2—2

of it. Call these A and B. A exerts a force on B in the direction
from B to A, and B exerts a force on A in the direction from A to
B. These two forces are merely two aspects or views of the one
tension in the rope at P.

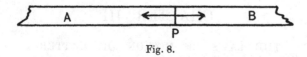

Fig. 8.

If a body is pushed along by a rigid rod there is an action
transmitted along the rod which is called a pressure and is of the
opposite kind to a tension. Consider, as with the rope, two parts
of the rod on either side of a cross section Q (Fig. 9). Call them

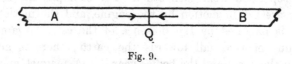

Fig. 9.

A and B. A pushes B in the direction from A to B. Also B
pushes A in the direction from B to A. The forces in a pressure
are opposite to those in a tension.

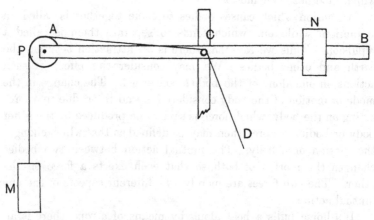

Fig. 10.

A simple experiment illustrating that when there is an action
between two bodies they are acted on by forces in opposite

directions is the following. A bar AB (Fig. 10) about 3 feet long
has a tube fixed to it at right angles at C. This tube is supported
by a fixed horizontal rod passing through it so that the bar can
turn about a horizontal axis at C. One end of the bar carries a
pulley P, and a sliding weight N can be fixed in any position near
its other end. A string is passed over the pulley from which a
mass M is suspended. The string is passed over the tube at
C and then down, and is held at D. The sliding weight is
adjusted so that the bar is balanced and will remain at rest in
a horizontal position. If now the string at D is suddenly pulled
down, this causes a tension in it which draws P and M together.
M moves up and P moves down, showing that the force on M is
upwards and that on P downwards.

If we bear in mind that a force is only one aspect of an action
between two or more bodies, we may go on to consider the
behaviour of a single body when acted on by a force. A force
changes the motion of a body, consequently if the motion of
a body remains unchanged we must conclude that there is no
force acting on it. This idea was expressed by Newton in his
first law of motion which may be stated as follows:—

Law I. Every body perseveres in its state of rest or of moving
uniformly in a straight line except in so far as it is made to
change that state by external forces.

A train running on a straight level track goes on with nearly
uniform velocity after the steam is shut off, and scarcely slows up
appreciably until the brakes are applied. What little slowing up
there is without the brakes can be explained as being due to the
resistance of the air and friction on the track. Anything which
diminishes the resistance to the motion of a body enables it to
continue longer in motion. The rotation of the earth goes on so
far as we know at a practically uniform rate, because there are no
appreciable forces tending to stop it.

If the velocity of a body is changing, that is, if it has an
acceleration, we say a force is acting on it and producing the
change. For example, it is found that any body falling freely has
an acceleration of approximately $980 \frac{\text{cms.}}{\text{sec.}^2}$ vertically downwards.

We say then that there is a force acting on it in this direction. This force is called its weight. It is one aspect of the mutual action between the body and the earth. In order to deal with forces scientifically it is necessary to define what we mean by the magnitude of a force and to adopt units in terms of which forces can be measured. As a preliminary to this we have to consider a number of experimental methods and results on which the plan adopted for the measurement of forces is based.

It has already been stated that the quantity of matter in a body is called its mass, but nothing has been said about how the quantities of matter in two bodies can be compared. A unit of mass is a certain carefully preserved piece of platinum. Such a unit is of no value unless the masses of other bodies can be expressed in terms of it. To make this possible we have to adopt some property of bodies as the measure of their masses. The practical method which is nearly always used for the comparison of quantities of matter consists in comparing their weights by weighing with a balance and a set of weights.

When a body is weighed on a balance the weights are adjusted till they balance the body. The weights are then attracted by the earth with a force equal to the force with which the earth attracts the body. If the body is weighed in another place the same weights are found to balance it, but this does not show that the force with which the earth attracts the body is the same as at the first place; it merely shows that the forces on the body at the two places are in the same ratio as the forces on the weights. It is found as a matter of fact that the force with which the earth attracts bodies does vary considerably from place to place. It is less near the equator than near the poles, and greater at sea level than at the top of a mountain.

A balance consists essentially of a rigid bar, called the beam, to which three knife edges are fixed. These knife edges are perpendicular to the length of the bar. When the bar is horizontal the middle knife edge faces downwards and the end edges upwards. The middle edge should be exactly half way between the other two. The middle edge rests on a horizontal plane and two pans are hung from planes which rest on the end edges. The balance is symmetrical about a vertical plane through the middle knife edge, and when nothing is in the pans the beam is in equilibrium when it is horizontal. If one pan is slightly pressed down and then let go, the beam oscillates about its equilibrium position. If two masses of equal weight are put in the pans, the equilibrium position of the beam remains unchanged. That is, if the force with which the attraction of the earth causes the masses in one pan to press down on the pan is equal to the force with which the other mass presses down on the other pan, then the equilibrium position of the beam is not altered. This follows from the symmetrical construction of the balance. By means of a balance, then, we can adjust the quantity of matter in a body or set of bodies so that its weight is equal to the weight of another body. Suppose we have a standard kilogram weight which has been compared with the standard kilogram by means of a balance. Then we can take a piece of brass weighing a little more than one kilogram and by carefully filing off small quantities of the brass we can make it of equal weight to the standard. We can then make a weight which balances the two kilogram weights together, which is therefore a two kilo weight. If another two kilo weight is made, we can make a five kilo weight by balancing it against the three weights; and so on. Also if we make two equal weights which together balance the standard kilogram, we know that they are each 500 grams, and if we make 1000 exactly equal weights which together balance the standard then we know that each is a one gram weight. In this way it is possible to make a set of weights having known values in terms of the standard. Such sets usually consist of weights of 1, 2, 2, 5, 10, 20, 20, 50, 100, 200, 200, 500 and 1000 grams. With a balance and a set of weights we can find the weight of any body by balancing it against the weights. The weight so found is the number of grams or pounds which, at the place where the weighing was done, is attracted by the earth with a force equal to the force with which the body weighed is attracted.

It is found that the weight of a body measured in this way by balancing it with a set of weights is the same wherever the weighing is performed, for example it is the same in Europe as in America and the same at the sea level as on the top of any mountain. Also it is not affected by the physical condition of the body, for example it is the same whether the body is hot or cold. It is found that the sum of the weights of two or more bodies is equal to the weight required to balance them when they are all put together in one pan of the balance. Also if a portion of a body is removed, then the weight of the body is diminished by an amount equal to the weight of the part removed. It has also been found as the result of many very careful experiments that the total weight of the matter in a system as determined by weighing with a balance and weights remains invariable so long as no matter either enters or leaves the system. This is true whatever processes, chemical, physical or biological, take place in the system. That the total weight of the matter in a system as determined by weighing with a balance and a set of weights remains constant is one of the most firmly established laws of nature. The accuracy with which weighings on a balance can be carried out is greater than that of any other measurement. It is possible to weigh a body of say 500 grams weight to one part in a million without very great difficulty.

Since it is found that the weight of any quantity of matter as measured with a balance and a set of weights remains invariable, it is natural to adopt the weight of a body as found with a balance and a set of weights as the measure of its mass or of the quantity of matter in it. We may therefore provisionally adopt the convention which is in accordance with universal practice that the quantities of matter in bodies are to be reckoned proportional to their weights as determined with a balance and a set of weights. We shall see later that the adoption of this convention can be further justified.

The unit of weight employed when the weight of a body is found with a balance and a set of weights is the weight of the unit mass at the place where the weighing is done. The numerical value of the mass of a body is therefore equal to the numerical value of its weight as found with a balance and a set of weights. Thus for example if we weigh a body with a balance and a set of weights and find that it weighs 20 pounds, then its mass is 20 pounds, or if it weighs 125 grams then its mass is 125 grams. A balance and a set of weights therefore enable the mass of a body to be found in terms of the unit of mass.

We may now go on to consider the motion of bodies under the action of forces. If a body moves with an acceleration, a force is said to be acting on it. It can be shown experimentally that the acceleration imparted to a body by a force is proportional to the force. To do this we may use a set of weights to give us forces having known ratios. For example, the force with which the earth attracts a five gram weight is five times that with which it attracts a one gram weight at the same place.

A convenient instrument for making experiments on the motion of bodies due to forces is known as Atwood's machine.

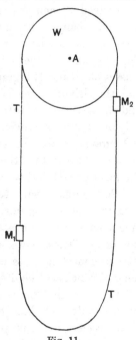

This machine is only suitable for rather roughly illustrating the laws of motion. Exact methods of verifying them will be described in later chapters. Atwood's machine consists of a light wheel W (Fig. 11) mounted on a horizontal axle A with a thread TT carrying two masses M_1 and M_2 passed over it. If the two masses are equal their weights balance each other, and they will remain at rest in any position. If they are set in motion up or down, they move with uniform velocity in accordance with the first law of motion. Suppose now that a small additional mass, say one gram, is put on the top of the mass M_2. The weight of this additional mass will cause M_2 to move down and M_1 up at an equal rate. If the times they take to move known distances starting from rest are measured, the distances are found to be proportional to the squares of the times. This shows that

Fig. 11.

the masses move with a uniform acceleration, for we have $s = \frac{1}{2} at^2$, where s is the distance described in a time t by a body starting from rest and moving with a uniform acceleration a. The acceleration is equal to $2s/t^2$, and so can be calculated from the observed distances and times. Thus it appears that the weight of one gram which is a constant force acting on the

Atwood's machine gives it a uniform acceleration. If now we take away one gram from M_1 and add it to M_2, the moving force will be the weight of three grams and the total mass moved will be unchanged. It will then be found that the masses move with three times their previous acceleration. If we take another gram from M_1 and put it on M_2 we get five times the acceleration due to the one gram weight. Thus it appears that the acceleration imparted to the machine is proportional to the force driving it. This is found to be true for all bodies. For any body then the ratio force ÷ acceleration has a definite value.

Instead of changing the force while the mass is kept constant we can change the mass while the force is kept constant. The masses can be found with a balance and a set of weights. Suppose we find the acceleration of the Atwood's machine with masses of 99 and 101 grams. The mass moved then is 200 grams and the driving force is the weight of two grams. If we next find the acceleration with masses of 199 and 201 grams it will be only one-half that previously obtained. With masses of 499 and 501 grams it will be only one-fifth. Thus when the driving force is kept constant the acceleration is inversely proportional to the mass moved. If we find the acceleration with masses of 198 and 202 grams or with 495 and 505 grams, it will be the same as with 99 and 101 grams. This shows that the force required to give a body a certain acceleration is proportional to the mass of the body. These laws which can be roughly proved true with Atwood's machine have been verified by many exact experiments, some of which will be described in later chapters.

The facts that the ratio force/acceleration has a definite value for any body and that the force required to impart to a body a given acceleration is proportional to its mass show that

$$\frac{f}{a} = cm,$$

where f denotes the force acting on a body, m the mass of the body, a the acceleration imparted to the body by the force f, and c is a constant.

By means of this equation we can define the unit of force. The units of acceleration and of mass have already been defined, so that the value of the unit of force is determined by the value of the constant c. Thus if $m = 1$ and $a = 1$ we get $f = c$. The unit of force used in scientific work is chosen so as to make $c = 1$, so that

$$\frac{f}{a} = m \quad \text{or} \quad f = ma.$$

If then $m = 1$ and $a = 1$ we get $f = 1$. The unit force therefore is the force which gives unit mass unit acceleration. This unit of force may be denoted by MLT^{-2}, where M, L and T denote as before the fundamental units of mass, length and time. It is called the dynamical unit of force.

If the unit of mass is the gram, the unit of length the centimetre, and the unit of time the second, the corresponding unit of force is called a *dyne*. A dyne is a force which gives a mass of one gram an acceleration of one cm. per sec. per sec.

The property of matter measured by the ratio f/a is sometimes called inertia. We have seen that it is found to be proportional to mass as measured by a balance and a set of weights. It is that property in virtue of which force is required to change the motion of matter. We know little more about force and inertia than the experimental fact that f/a is constant for any piece of matter. The ideas of force and inertia cannot be explained; they are fundamental in mechanics like the ideas of space and time, and are acquired by experience. Anyone who has had to deal with large masses in motion, such as motor cars, railway trains or heavy fly-wheels, knows the difficulty of stopping or starting them quickly and has acquired the ideas of force and inertia by experience. In the scientific study of mechanics we endeavour to make our ideas as precise as possible and to develop methods of accurately measuring such quantities as space, time, force and mass, but we need not attempt to explain their real nature; we are in fact unable to do so.

For a body moving along a straight line with a uniform acceleration a we have

$$a = \frac{v_2 - v_1}{t},$$

where v_1 is its velocity at the beginning of an interval of time t and v_2 its velocity at the end. Substituting this in the equation $f = ma$ we get

$$f = \frac{mv_2 - mv_1}{t}.$$

The product of the mass of a body and its velocity is called its *momentum*. Thus the force acting on the body is equal to the increase in the momentum of the body in unit time or to the rate of increase of the momentum with time. Newton expressed the relations between force and momentum in his second law of motion which may be stated as follows:—

Law II. The change of momentum of a body is proportional to the force acting on it and to the time during which the force acts and is in the same direction as the force.

This agrees with the equation $m(v_2 - v_1) = ft$ which is equivalent to $f = ma$.

The mutual action between two bodies A and B can be looked upon as made up of the force which A exerts on B and the force which B exerts on A. These two forces are the different aspects of the mutual action. They are sometimes referred to as the action and reaction.

Newton's third law of motion may be stated as follows:—

Law III. The action and reaction between two bodies are equal and in opposite directions.

It is important to remember that the action and reaction act on different bodies. The action is the force exerted by A on B and the reaction the force exerted in the opposite direction by B on A. Consider the case of a horse pulling a boat by means of a rope. Let P and Q (Fig. 12) be two cross sections of the rope.

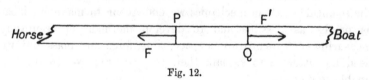

Fig. 12.

Consider the forces acting on the portion of the rope between P and Q. At P the rope on the left is pulling PQ towards the horse with a force F, say. At Q the rope to the right of Q is pulling PQ towards the boat with a force F', say. The resultant force on the portion of the rope between P and Q is therefore $F - F'$ in the direction of the horse. Let the mass of the rope between P and Q be m. Then we have

$$F - F' = ma,$$

where a is the acceleration with which the rope is moving to the left. Now suppose that P and Q are taken close together

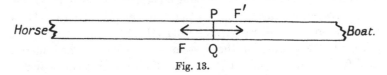

Fig. 13.

(Fig. 13) so that there is no mass between them. Then $m = 0$ so that

$$F - F' = 0.$$

When P and Q coincide F and F' become the two forces at PQ which make up the tension in the rope at that cross section. The equation $F - F' = 0$ just obtained shows that F and F' are equal and they are in opposite directions. F and F' are the action and reaction at the cross section PQ of the rope.

When one body acts on another the action and reaction across any surface separating them may be seen to be necessarily equal and opposite in the same way as at the section of the rope just considered.

In the case of a mutual action between bodies which are not in contact, such as the attraction between the earth and a body near it or the attraction between a piece of iron and a magnet, the action is transmitted through a medium called the ether which fills all space. The ether enables the magnet and the piece of iron to act on each other in some way not well understood, so that it takes the place of the rope in the case of the horse and boat. The action and reaction at any cross section in the ether are equal and opposite just as in the rope.

When a horse pulls a cart along a level road the backward force exerted by the cart on the horse is equal and opposite to the forward force exerted by the horse on the cart. The forces acting on the cart are this forward force due to the horse and retarding forces due to friction. If the forward pull is greater than the retarding forces, the cart moves forward with an acceleration. The forces acting on the horse are the backward pull of the cart and forward forces due to the reaction of the road when the horse presses on it with its hoofs. If the forward forces due to the exertions of the horse are greater than the backward pull of the cart, the horse moves forward with an acceleration.

In any system of bodies the mutual actions between the bodies in the system are made up of equal and opposite forces. These internal forces therefore balance each other so that they have no effect on the motion of the system as a whole. They merely modify the arrangement of its parts. Forces acting on bodies in the system due to mutual actions between bodies outside the system and bodies in it are called external forces and modify the motion of the system as a whole.

In the case of the horse and cart on a level road the action between them is made up of two internal forces and so does not affect the motion of the horse and cart considered as a whole. The external forces are the frictional resistances retarding the cart and the forward forces due to the reaction of the road on the horse's hoofs. When the forward forces are greater than the retarding friction the horse and cart move forward with an acceleration. If the forward forces due to the exertions of the horse are equal to the backward forces due to friction, the horse and cart either remain at rest or, if moving, continue to move along with a constant velocity.

Consider two bodies A and B, and suppose that there is a mutual action between them, a repulsion say. There will be a force on B in the direction from A to B which will increase the momentum of B in that direction. The rate of increase of the momentum of B is equal to the force. There will also be an equal and opposite force on A which will increase the momentum of A in the direction from B to A and the rate of increase of the momentum of A in this direction will be equal to the rate of increase of the momentum of B in the opposite direction. If we regard momentum in the direction from A to B as positive and that in the opposite direction as negative, then A gains positive momentum and B gains negative momentum at an equal rate. Thus the total algebraical change of momentum is zero and the momentum gained by A is equal to that lost by B. A mutual action between two bodies may therefore be said to consist of a transference of momentum from one to the other. The total momentum remains constant. A repulsion produces a flow of momentum in the positive direction and, as may easily be seen, an attraction produces a flow in the negative direction.

When the mutual action between two bodies lasts only for a very short time, as in a collision between two billiard balls, it is often convenient to measure the amount of the action by the amount of momentum transferred from one body to the other. Each body is said to be acted on by an impulse which is taken equal to the change in the momentum of the body. The impulse received by one body is equal and opposite to that received by the other.

If two masses m_1 and m_2, both moving along the same straight line with velocities v_1 and v_2, collide, and after the collision continue to move along the same straight line as before with velocities v_1' and v_2', then we have

$$m_1 v_1 + m_2 v_2 = m_1 v_1' + m_2 v_2'.$$

If the bodies stick together when they collide, then $v_1' = v_2' = v$, say,

and $$m_1 v_1 + m_2 v_2 = (m_1 + m_2) v.$$

If velocities in one direction are counted positive, then those in the opposite direction along the line are counted negative.

For example, suppose two bodies with masses of 10 pounds and 30 pounds are moving along a straight line. Let the velocity of the 10 pound mass be 5 feet per second and that of the 30 pound mass be 10 feet per second in the opposite direction. After the collision let them stick together and move with a velocity v. Then we have

$$v_1 = +5 \text{ and } v_2 = -10,$$

so that $$(10 \times 5) - (30 \times 10) = 40v,$$

or $$v = -6\tfrac{1}{4} \text{ feet per second.}$$

The momentum lost by the 10 pound mass is $10 (5 + 6.25) = 112.5$, and that gained by the 30 pound mass is $30 (10 - 6.25) = 112.5$.

If the bodies do not stick together but rebound from each other, then it is found that the relative velocity after impact bears a constant ratio, less than one, to the relative velocity before impact and is in the opposite direction. Hence

$$\frac{v_1' - v_2'}{v_2 - v_1} = e,$$

where e is a constant less than one. The value of e depends on the nature of the bodies. The constant e is called the coefficient of restitution.

The most important kind of force is the force with which the earth attracts bodies. This force is called weight. As we have seen, the weight of a body is usually stated in terms of the weight of unit mass at the same place. It is then numerically equal to the mass of the body. The weight of unit mass at any place is often used as a convenient unit of force, but as a unit of force it

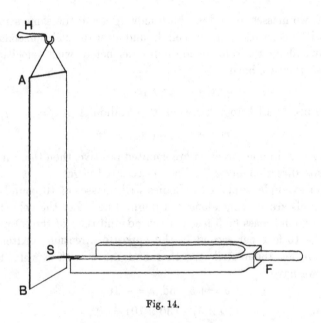

Fig. 14.

has the serious drawback that it is not of exactly the same value in different places. In dealing with the weights of bodies we are usually concerned with their weights as a measure of the quantities of matter in them. In such cases the weight is expressed in terms of the weight of unit mass. When the weight of a body is employed to set other bodies in motion or to balance forces, we are concerned with it as a force, and then it is best to express it in terms of a dynamical unit of force, such as the dyne, which always has the same value. It is important to bear in mind these two ways of using weight either as a measure of mass or as a force. Weight of course is a force in all cases.

If a number of balls made of different substances, say cork, wood, iron and lead, are put in a box and the box is turned upside down and then opened so that they all fall out together, it is found that all reach the ground at the same time. Very light bodies like feathers or pieces of paper fall more slowly; but this can be shown to be due to the resistance of the air. If a feather and a piece of lead are dropped together in a vacuum, they fall at the same rate.

The way in which a falling body moves can be examined in the following way:—

A glass plate AB (Fig. 14) is hung vertically by a thread from a hook H. The plate has been coated with lamp black by holding

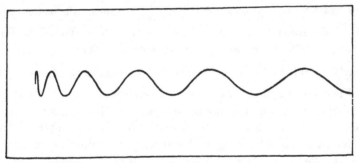

Fig. 15.

it in a smoky flame. A tuning fork F is supported so that the tip of a light pointer S attached to one prong touches the plate near the bottom. The tuning fork is set vibrating, and then the thread is burned with a match so that the plate falls. As it moves past the vibrating fork the pointer carried by the fork traces a wavy line in the lamp black on the plate. The appearance of such a line is shown in Fig. 15. If the fork makes, say, 200 vibrations in a second, then the distance between the crests of the waves will be the distance travelled by the plate in successive $\frac{1}{200}$ths of a second. If we make a mark on the plate at every ten waves and then measure the distances of these marks from the beginning of the wavy line, we shall have the distances travelled in $\frac{1}{20}, \frac{2}{20}, \frac{3}{20}, \frac{4}{20}$

seconds, and so on. The following table gives results which might have been obtained in this way.

Time (t)	Distance (s)	$2s/t^2$
0·05 sec.	1·23 cm.	984
0·10	4·90	980
0·15	11·03	981
0·20	19·60	980
0·25	30·63	980

The third column gives the values of $2s/t^2$, which are practically constant and equal to 980. Thus it is found that the distances fallen by the plate are proportional to the squares of the times from the start. We have seen that when a body starts from rest and moves with a uniform acceleration a the space s described in a time t is given by $s = \frac{1}{2}at^2$. Hence $a = 2s/t^2$. Thus the experiment shows that the plate falls with a uniform acceleration nearly equal to $980 \frac{\text{cms.}}{\text{sec.}^2}$. Since all bodies fall equal distances from rest in equal times when allowed to fall freely, it follows that all bodies fall with a uniform acceleration nearly equal to $980 \frac{\text{cms.}}{\text{sec.}^2}$. This acceleration is usually denoted by g and is called the acceleration of gravity. More exact methods of finding its value and proving more accurately that it has the same value for all substances will be described in a later chapter. It is found that the value of g is not exactly the same at different places. For example, it is less at the top of a mountain than at sea level; but at any particular place it is equal for all substances.

It was stated earlier in this chapter that the adoption of the weight of a body, as measured with a balance and a set of weights, as a measure of its mass would be further justified. Theoretically it may be objected that this way of measuring the mass of a body depends too much on things accidentally present—like the earth, and the balance. It ought to be possible to formulate a definition of the mass of a body in terms of properties of the body and independent of the presence of other bodies. This can be done by means of the property of inertia, which is measured by the ratio of force to acceleration or $f \div a$. We define the mass m of a body as proportional to its inertia, so that

$$\frac{f}{a} = cm,$$

where c is a constant. We then fix the size of the unit of force by
putting $c = 1$ and so get $f = ma$ as before.

If w denotes the weight of a body expressed in dynamical
units of force, then since it gives the body an acceleration g we
have $w = mg$. But at any particular place g is the same for all
bodies, hence w is proportional to m. If then the weight is
expressed in terms of the weight of unit mass, it is numerically
equal to the mass. It follows that the weight as found with a
balance and a set of weights is numerically equal to the mass
when the mass is defined as proportional to the inertia. All the
experimental results described earlier in this chapter, proving
weight as measured with a balance and a set of weights, that is
when expressed in terms of the weight of unit mass, to be
invariable, apply therefore to inertia; so that we have the same
reasons for adopting inertia as the measure of the quantity of
matter as we had for adopting the weight measured with a
balance and a set of weights. Inertia has the advantage that it is
a property of the body independent of the presence of other bodies.
This is a purely theoretical advantage and in practice masses are
almost always determined by weighing with a balance and a set of
weights. Since g is the same for all bodies at the same place, the
balance method gives the same results as would be obtained if the
masses were compared by finding their accelerations due to known
forces. The fact that the total mass of any system remains
invariable, so long as no matter enters or leaves it, is often referred
to as the principle of the conservation of matter or of mass.

The equation $w = mg$ gives the weight w of a mass m in
dynamical units of force. If m is in grams and g in cms. and secs.,
then w is in dynes. The weight of a gram is therefore g dynes or
nearly 980 dynes. If m is expressed in pounds and g in feet
and seconds, we get the weight in terms of the force which gives a
one pound mass an acceleration of one foot per sec. per sec. This
unit of force is called a *poundal*. Since g is about $32 \frac{\text{feet}}{\text{sec.}^2}$, the weight
of a pound is equal to about 32 poundals. Since the value of g is
different at different places it follows that the number of dynamical
units of force in the weight of a body is different at different
places. The value of g at sea level and latitude λ is nearly equal

to $978 \left(1 + 0\cdot00531 \sin^2 \lambda\right) \dfrac{\text{cms.}}{\text{sec.}^2}$. At the equator this gives

$$g = 978 \frac{\text{cms.}}{\text{sec.}^2},$$

and at the north pole, where $\lambda = 90°$,

$$g = 983\cdot19 \frac{\text{cms.}}{\text{sec.}^2}.$$

In practical work a force equal to the weight of a unit of mass is often used as a convenient unit of force. For example, a pound weight is the unit of force used by English and American engineers. If a force f gives a body whose weight is w units of force an acceleration a, then since its weight would give the body an acceleration g we have

$$\frac{f}{w} = \frac{a}{g}$$

or

$$f = w \frac{a}{g}.$$

This equation is true whatever unit of force is used, so long as both f and w are expressed in terms of the same unit. If w is given in dynes we get f in dynes, and if w is given in pounds weight we get f in pounds weight. For example, if a body whose weight is 1960 dynes moves with an acceleration $10 \dfrac{\text{cms.}}{\text{sec.}^2}$ the force acting on it is $1960 \times \dfrac{10}{980} = 20$ dynes. If the weight of the body is taken to be 2 grams weight we get the force equal to $2 \times \dfrac{10}{980} = 0\cdot02041$ grams weight. If a force of 10 pounds weight acts on a body whose weight is 100 pounds, the acceleration produced is

$$\frac{10}{100} \times 980 = 98 \frac{\text{cms.}}{\text{sec.}^2} \text{ or } \frac{10}{100} \times 32 = 3\cdot2 \frac{\text{feet}}{\text{sec.}^2}.$$

The equation $f = w \dfrac{a}{g}$ is a very convenient form of the fundamental equation and may often be used in working out practical problems. a and g must both be expressed in terms of the same unit and f and w also. For example, if a kilogram weight moves with an acceleration of 16 $\dfrac{\text{feet}}{\text{sec.}^2}$ then the force acting on it is

given by $f = 1000 \times \dfrac{16}{32} = 500$ grams weight.

If a body weighing 10 pounds moves with an acceleration of 490 $\dfrac{\text{cms.}}{\text{sec.}^2}$ the force on it is given by $f = 10 \times \dfrac{490}{980} = 5$ pounds weight. If the force is wanted in poundals then the weight of 10 pounds is put equal to 320 poundals so that

$$f = 320 \times \dfrac{490}{980} = 160 \text{ poundals.}$$

CHAPTER IV

FORCE AND MOTION

IF two forces act on a particle of matter, each imparts to it the

Composition of Forces. acceleration given by the equation $f = ma$. The resulting motion can be got by adding together the two motions due to each force acting alone. Let P (Fig. 16) be a particle of mass m and let a force f act on it in the direction PA. Let

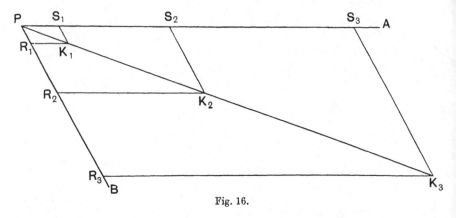

Fig. 16.

PS_1, PS_2, PS_3 be the distances it would travel along PA in times τ, 2τ and 3τ starting from rest at P if acted on by f alone.

These distances are given by the equation $s = \frac{1}{2}at^2$ and are therefore proportional to 1, 4, 9 respectively. Let another force f' act on the particle in the direction PB and let PR_1, PR_2, and PR_3 be the distances it would describe in the intervals τ, 2τ, and 3τ due to the action of this force. If both forces act together in the time τ the particle will move a distance PS_1 in the direction PA and a distance PR_1 in the direction PB. It will therefore arrive at K_1 the opposite corner of the parallelogram $PS_1K_1R_1$. In the

same way after a time 2τ it will be at K_2 and after 3τ at K_3. Thus
it appears that it moves along $PK_1K_2K_3$. Also since PK_1, PK_2
and PK_3 are proportional to PS_1, PS_2 and PS_3 it follows that the
distances described along PK_3 are proportional to the squares of the
times taken to describe them. The particle therefore moves along
PK_3 with a uniform acceleration. Since $s = \tfrac{1}{2}at^2$, the distance
described in any time is proportional to the acceleration, and the
acceleration is proportional to the force. The distances PS_3 and
PR_3 may therefore be taken to represent the forces f and f' in
magnitude and direction, and in the same way PK_3 will represent
a single force which would make the particle move along PK_3 in the
same way as it does under the action of f and f'. This force

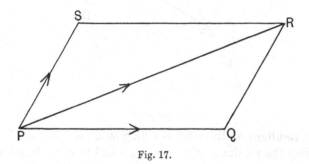

Fig. 17.

represented by PK_3 is called the resultant of the forces f and f',
which are represented by PS_3 and PR_3. The resultant of two
forces acting on a particle can therefore be obtained by drawing
a line PQ representing one of them in magnitude and direction,
and then drawing another line QR representing the other; the
resultant is then represented by PR. If from P we draw PQ
representing one force and PS representing the other, then the
resultant is represented by the diagonal PR of the parallelogram
$PQRS$.

Since a force has magnitude and direction it is a vector, so the
method of finding the resultant of the two forces is the same as the
method of adding two vectors. The same is also true for accelerations,
for PS_3 and PR_3 are proportional to the accelerations along PA
and PB, and PK_3 is proportional to the resultant acceleration.

If PQ (Fig. 17) $= a$, $PS = b$ and the angle $QPS = \theta$, then

$$PR^2 = a^2 + b^2 + 2ab \cos \theta,$$

for

$$PR^2 = (PS \sin \theta)^2 + (SR + PS \cos \theta)^2$$

or

$$PR^2 = b^2 \sin^2 \theta + a^2 + 2ab \cos \theta + b^2 \cos^2 \theta$$

and

$$b^2 \sin^2 \theta + b^2 \cos^2 \theta = b^2.$$

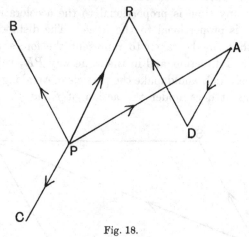

Fig. 18.

The resultant of three forces acting on a particle can be found by finding the resultant of two of them and then the resultant of this and the third. Thus suppose PA, PB and PC represent three

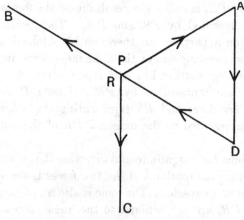

Fig. 19.

forces acting on a particle at P (Fig. 18). Draw AD equal and
parallel to PC, and DR equal and parallel to PB. Then PR
represents the resultant of the three forces.   In the same way the
resultant of any number of forces acting on a particle can be found.

If the resultant of any number of forces acting on a particle is
zero, then the forces are said to be in equilibrium.   For example, if
with three forces PA, PB and PC (Fig. 19) we draw AD equal
and parallel to PC and DR equal and parallel to PB and R falls
on P, then the three forces just balance each other and have no
resultant.   In this case the sides of the triangle PAD taken in
order represent the three forces in magnitude and direction.

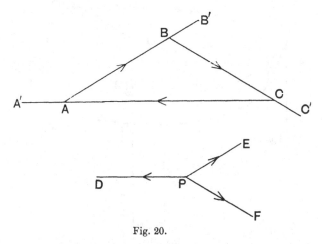

Fig. 20.

Let ABC (Fig. 20) be a triangle representing three forces in
equilibrium acting on a particle at P in directions PD, PE and PF,
which are parallel respectively to CA, AB and BC.   Produce

$$CA \text{ to } A', \quad AB \text{ to } B' \quad \text{and} \quad BC \text{ to } C'.$$

We have

$$A'\hat{A}B = D\hat{P}E, \quad B'\hat{B}C = E\hat{P}F \quad \text{and} \quad C'\hat{C}A = F\hat{P}D.$$

The sides AB, BC and CA, of the triangle ABC, are proportional
to the sines of the angles BCA, CAB and ABC.   These sines are
equal to the sines of the angles $C'CA$, $A'AB$ and $B'BC$.   Consequently, since the sides of the triangle ABC represent in
magnitude and direction the three forces acting at P, it follows that

the forces in the directions *PD, PE* and *PF* are proportional to the sines of the angles *EPF, DPF* and *EPD*; that is each force is proportional to the sine of the angle between the other two.

This can be verified with the apparatus shown in Fig. 21. *TT* is a horizontal circular table with its circumference graduated in degrees. Round it three pulleys *A, B, C* can be clamped in any positions. Threads passing over the pulleys carry weights W_1, W_2, W_3, and are tied together at *O*. By adjusting the weights and moving the pulleys round the table, *O* can be made to rest over the centre of the table. The angles between the threads can then be read off

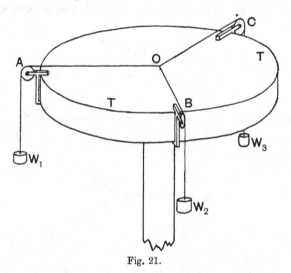

Fig. 21.

on the graduated circumference. The sines of the angles *COB*, *AOC* and *AOB* are found to be proportional to the weights W_1, W_2 and W_3. Since any two sides of a triangle are together greater than the third side, it follows that any two of three forces in equilibrium must be together not less than the third.

If two forces act on a particle in directions at right angles to each other, then the distance the particle moves in the direction of either force is the same as if the other were not acting. Let *PQ* (Fig. 22) represent a force acting at *P*. Take a line in any direction *PA* and drop a perpendicular *QN* from *Q* on to *PA*. The force *PQ* may be regarded as the resultant of the two forces represented by

PN and NQ. The force PN is called the resolved part of PQ along the direction PA. The perpendicular component NQ does not affect the motion in the direction of PA. If the angle QPA is denoted by θ, then we have $PN = PQ \cos \theta$ and $NQ = PQ \sin \theta$.

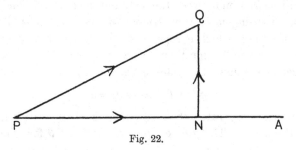

Fig. 22.

As an example consider the case of a body P resting on a smooth inclined plane ABC (Fig. 23).

Let PQ drawn vertically downwards represent the weight of the body. Draw PK parallel to AB and drop a perpendicular QN on it from Q. The component of the weight represented by NQ does not tend to move the body along the plane and is balanced by

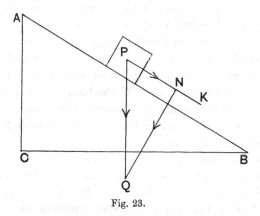

Fig. 23.

the equal and opposite reaction of the plane on the body. The component PN acts along the plane. If m denotes the mass of the body, then its weight is mg dynamical units of force and the component represented by PN is equal to $mg \dfrac{PN}{PQ}$. The angle ABC

is equal to the angle PQN. Denoting this angle by θ, we have for the component along the plane $mg \sin \theta$. If there were no friction the body would therefore move down the plane with an acceleration a given by the equation $mg \sin \theta = ma$ or $a = g \sin \theta$. Actually there will be friction between the body and the plane. This friction exerts a retarding force proportional to the force exerted by the body on the plane which is represented by QN and so is equal to $mg \cos \theta$. Let the friction then be $\mu mg \cos \theta$ where μ is a constant. The acceleration down the plane is then given by

$$g (\sin \theta - \mu \cos \theta) = a.$$

If the friction is greater than $mg \sin \theta$, the body will remain at rest on the plane. The greatest possible value of θ for which this can happen is given by

$$\sin \theta - \mu \cos \theta = 0$$

or $$\mu = \tan \theta.$$

μ is called the coefficient of friction. It depends on the state and nature of the surfaces in contact. It is diminished usually by coating the surfaces with oil.

We have seen in Chapter II that a point moving round a circle

Force on a body moving in a circle. of radius r with a uniform velocity v has an acceleration directed towards the centre of the circle and equal to v^2/r. If a particle of mass m moves round a circle, a force equal to mv^2/r directed towards the centre is therefore required to keep it on the circular path. If the body makes n revolutions per second we have $v = 2\pi rn$ so that the force required is equal to $4\pi^2 mn^2 r$. This force may be measured approximately with the apparatus shown in Fig. 24.

A frame F is mounted on a vertical axle and can be made to rotate by means of a belt and pulley as shown. The frame carries a horizontal rod BB' on which a mass M can slide freely. Two cords attached to this mass pass round pulleys at P attached to the rod, and then go up along the axis of rotation to a spring balance S. The mass M is balanced by an equal mass M', which can be fixed at any distance from the centre of the rod.

When the apparatus is rotating, the mass M moves towards B until the tension in the strings is sufficient to keep it from moving further. The tension in the strings is indicated by the balance. With this apparatus the force on the mass M, the radius of its circular path and the number of revolutions per second can be easily found and the force can be compared with the theoretical value

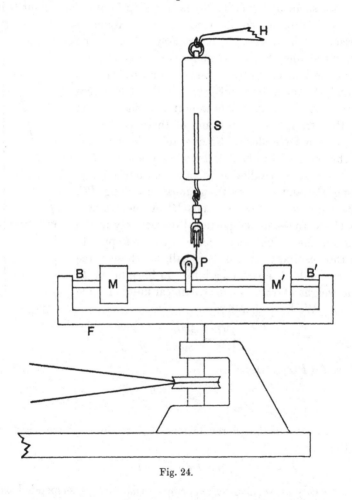

Fig. 24.

$4\pi^2 m n^2 r$. The tension in the strings, it should be observed, pulls the mass M in towards the centre and also pulls the centre outwards with an equal force. These two equal and opposite forces are an example of an action and reaction.

When a wheel is rotating, the rim is kept on its circular path by a tension in the spokes of the wheel. If the spokes are not strong enough the wheel may break in pieces, and then its parts fly out along tangents to their circular paths.

If a small ball P (Fig. 25) is hung up from a fixed point O by a fine thread, it can be made to move round a circular path by starting it properly. Let OA be a vertical line through O and PC a perpendicular from the ball on to OA. Suppose the ball moves round OA in a circle of radius CP. The forces acting on the ball are its weight and the tension in the string. The resultant of these two forces must be a force along PC equal to mv^2/r, where m is the mass of the ball, v its velocity and $r = CP$. The weight is parallel to OC and the tension acts along PO, and their resultant must act along PC. The three sides PO, OC and CP of the triangle POC are therefore proportional respectively to the tension, the weight and a force equal and opposite to the resultant force on the ball, for these three forces would keep the ball in equilibrium at P. The weight of the ball in dynamical units of force is mg. We have therefore

Fig. 25.

$$\frac{PC}{OC} = \frac{mv^2/r}{mg} = \frac{r}{\sqrt{l^2 - r^2}},$$

where $l = PO$.   Hence

$$v^2 = \frac{r^2 g}{\sqrt{l^2 - r^2}}.$$

If the ball makes n revolutions per second we have $v = 2\pi rn$ so that

$$g = 4\pi^2 n^2 \sqrt{l^2 - r^2}.$$

It can easily be verified by experiment that such a suspended ball moving round circles makes the number of revolutions per second given by this formula. This shows that the expression mv^2/r or $4\pi^2 mrn^2$ for the force required to keep the ball on its circular path is correct.

Let us now consider the case of a body moving with a simple

Force on a body moving with a simple harmonic motion.

harmonic motion. We have seen in Chapter II that a simple harmonic motion is an oscillation backwards and forwards along a straight line, such that the acceleration is proportional to the distance from the middle point of the oscillation and directed towards it. If a body of mass m moves in this way there must be a force equal to ma acting on it, which force is therefore proportional to its distance from the middle point of the oscillation and directed towards this point. Let this force be equal to $-cx$, where x is the displacement of the body from the middle point and c is a constant. The force is taken equal to $-cx$ not $+cx$ because the force is in the opposite direction to the displacement, so that when x is in the positive direction the force is in the negative direction. The acceleration a is then given by the equation

$$a = -\frac{cx}{m}.$$

We found in Chapter II that the time of a complete vibration is given by $T = 2\pi/\sqrt{\mu}$ when the acceleration is equal to $-\mu x$. We have therefore $\mu = c/m$, so that the time of vibration of the body is given by the equation

$$T = 2\pi\sqrt{\frac{m}{c}}.$$

Consider the case of a mass hung up on a spiral spring (Fig. 26). The weight of the mass will pull out the spring until the upward force exerted by the spring is equal to the weight. It is found that the extension of the spring is proportional to the weight of the body hung from it. The upward force due to the spring may therefore be put equal to $-ce$, where c is a constant and e the extension. If a mass m is hung from the spring, the mass will rest in equilibrium when the extension is e' and $ce' = mg$.

If now the mass is pulled down a distance x below its equilibrium position, there will be a resultant force on it given by

$$-c(e' + x) + mg = -cx.$$

Thus the resultant force on the mass due to the spring and to its weight is proportional to the distance x of the mass from its position of equilibrium. If the mass is pulled down and then let go, it therefore oscillates up and down in a simple harmonic motion of period $2\pi\sqrt{\dfrac{m}{c}}$ which is independent of its amplitude. The spring oscillates up and down with the mass, and it can be shown that one-third of the mass of the spring should be added to the mass of the body when calculating the period by means of the formula just found.

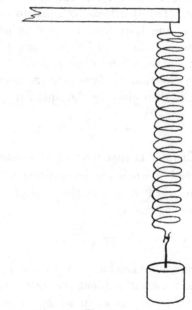

Fig. 26.

Now consider what is called a simple pendulum. This consists of a small heavy sphere hung from a fixed point by a fine thread. If the sphere is pulled to one side and let go it oscillates backwards and forwards.

Simple Pendulum.

Let O (Fig. 27) be the fixed point and P the small sphere of mass m. Let $OP = l$. Draw a vertical line OC through O and a circle PCQ with centre O. Then P oscillates along PCQ and C is its equilibrium position. The tension in the thread has no component along the arc of the circle at P, so that it does not affect the motion of P along the arc. Draw a tangent at P meeting OC produced at A. The weight of P acts parallel to OC so that its component along PA is equal to

$$mg \cos P\hat{A}O \quad \text{or} \quad mg \sin P\hat{O}C.$$

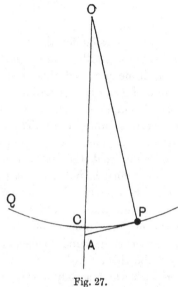

Fig. 27.

Let the length of the arc CP be denoted by x. Then

$$P\hat{O}C = \frac{x}{l},$$

so that the acceleration of P along the arc is equal to

$$-g \sin \left(\frac{x}{l}\right).$$

If x is very small we may put

$$\sin \left(\frac{x}{l}\right) = \left(\frac{x}{l}\right),$$

so that the acceleration of P along the arc is equal to $-g \dfrac{x}{l}$, and so is proportional to the displacement of P from C and in the opposite direction. It appears therefore that P will move with a simple harmonic motion along the arc provided its amplitude of vibration is very small. The time of a complete oscillation is given by

$$T = 2\pi \sqrt{\frac{m}{c}} = 2\pi \sqrt{\frac{l}{g}},$$

since $c = \dfrac{mg}{l}.$

It is easy to verify this expression for the time of oscillation of a simple pendulum experimentally, and to show that it is independent of the amplitude provided this is small. If l and T are measured, the value of g can be calculated.

Another example of simple harmonic motion is the motion of the prongs of a vibrating tuning fork. The restoring force is due to the elasticity of the prongs and is proportional to their displacements. It is found that the number of vibrations per second made by a tuning fork is independent of the amplitude of vibration, as it should be for a simple harmonic motion.

The fact that in cases like those just considered, where a body is acted on by a restoring force proportional to its displacement from a fixed point, the period is found to be independent of the amplitude, as it should be theoretically, is the best proof we have that the acceleration with which a body moves is proportional to the force acting on it, for the periods of such vibrations can be measured with great accuracy.

REFERENCES

Mechanics, Cox.
Experimental Mechanics, Sir R. Ball.

CHAPTER V

WORK AND ENERGY

WHEN a force acts on a body and the body moves in the
direction of the force, the force is said to do work.
The work done is equal to the product of the
force into the distance through which it acts. If w denotes the
work done, f the force and s the distance, then $w = fs$.

If $f = 1$ and $s = 1$ then $w = 1$, so that the unit of work is
the work done by a unit force when it acts through unit distance.
If the unit force is a dyne and the unit distance a centimetre,
then the unit of work is called an erg. The unit of work may be
denoted by

$$MLT^{-2} \times L \quad \text{or} \quad ML^2T^{-2},$$

where M, L and T respectively denote the fundamental units of
mass, length and time.

Let AB (Fig. 28) represent a force f, acting at B, and suppose

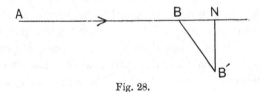

Fig. 28.

B is displaced to B'. Draw $B'N$ perpendicular to AB produced.
The work done by the force f is $f \times BN$ not $f \times BB'$. If BB'
were perpendicular to AB then no work would be done by f.

The most familiar example of work is that done in raising
weights to a higher level. The weight of a body acts vertically
downwards, so the work done in raising a body is equal to its
weight multiplied by the vertical height through which it is

4—2

raised. For example, if a body of mass m is pushed up an inclined plane of length l and height h, the work done against weight is mgh and so is independent of l. If the inclination of the plane to the horizontal is θ, then the component of the weight down the plane is $mg \sin \theta$, so that the work done in pushing the body up is $mg \sin \theta \times l$. But $l \sin \theta = h$ so that

$$mg \sin \theta \times l = mgh.$$

In calculating work we may therefore either multiply the force by the component of the displacement in the direction of the force or multiply the displacement by the component of the force in the direction of the displacement.

Forces are sometimes expressed in terms of the weight of unit mass instead of in terms of the force required to give unit mass unit acceleration. Thus we may say a force is equal to the weight of ten pounds. This way of measuring forces is often convenient. It is not suitable for very exact work because, as we have seen, the weight of unit mass varies slightly from place to place. If the force is measured in terms of the weight of a pound as unit and the distance through which it acts is expressed in feet, then the work is said to be in *foot-pounds*. A foot-pound is an amount of work equal to the work done in raising one pound a foot high. The foot-pound is a convenient unit of work often used by engineers.

As another example of work consider the work done in stretching a spiral spring. Let l_0 be the length of the spring when not stretched and let l' be the length when pulled out by a force f. The force required to stretch a spring is proportional to the increase of length of the spring, so that the force required increases uniformly from zero when the length is l_0 to f when it is l'. The average value of the force over the distance $l' - l_0$ is therefore $\frac{1}{2}f$, so that we have

$$w = \tfrac{1}{2} f (l' - l_0),$$

where w denotes the work done. This result may be obtained in another way as follows. In Fig. 29 let the horizontal distance ON_1 measured along OA from O represent the extension $l_1 - l_0$ of the spring and the vertical distance N_1P_1 above the line OA

the corresponding force required to stretch the spring from the length l_0 to l_1.

If a series of such points P_1, P_2, P_3, P_4 etc. are marked, they will all fall on a straight line through O, because the force required

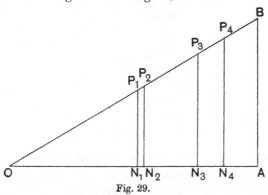

Fig. 29.

is proportional to the extension. Take two such points P_1 and P_2 very near together, and consider the work done in stretching the spring from N_1 to N_2. The force at N_1 is N_1P_1 and at N_2 it is N_2P_2. The work is therefore between $N_1N_2 \times N_1P_1$ and $N_1N_2 \times N_2P_2$. The area $N_1P_1P_2N_2$ also lies between $N_1N_2 \times N_1P_1$ and $N_1N_2 \times N_2P_2$. The difference between these limits becomes negligible if N_1N_2 is made very small, so that we see that the work is represented by the area $N_1P_1P_2N_2$. The area between OB and OA can be divided into a great many very narrow vertical strips like $N_1P_1P_2N_2$, the area of each one of which can be shown in the same way as for $N_1P_1P_2N_2$ to represent work done in stretching the spring. Thus we see that the area $N_1P_1P_4N_4$ represents the work done in stretching the spring from N_1 to N_4. If OA represents the extension $l' - l_0$ due to a force f represented by AB, then the work done during this extension is represented by the area OBA which is equal to $\frac{1}{2}AB \times OA$ or to $\frac{1}{2}(l' - l_0)f$.

A diagram like the one just considered, which shows the relation between force and displacement and on which area represents work, is called an *indicator diagram*, and is often made use of in engineering work. Suppose on such a diagram the relation between force and displacement is represented by a closed curve like $ABCD$ (Fig. 30). When the displacement increases from OA' to OC' the work done is equal to the area $A'ABCC'$, but when the displacement diminishes from OC' back to OA' the

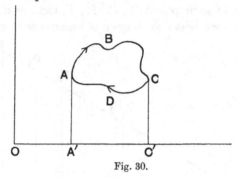

Fig. 30.

work done is negative and equal to the area $A'ADCC'$. The area $ABCD$ therefore represents the total work done during a displacement from A' to C' and then back to A'.

When an external force does work on a system of bodies the arrangement of the bodies in the system is changed. It is found that to bring the system back exactly to its original state, the system must be allowed to do an amount of work equal to the work done on it by the external force. Thus it appears that doing a definite amount of work on a system increases the system's power of doing work by an equal amount. The power of doing work possessed by a system is called its *energy*. Thus when work is done on any system by external forces the energy of the system is increased by an amount equal to the work done. The external system whose action provides the external forces loses an amount of energy equal to that gained by the other system. The total amount of energy therefore remains unchanged. In any system of bodies on which there are no external forces acting the total energy remains constant. This is a firmly established law of nature and is known as the principle of the conservation of energy.

For example, if a heavy body is moved up to a higher level, work is done on the system consisting of the body and the earth. If the body is allowed to move back to its original position, it does an amount of work equal to that done in raising it. The work done in stretching a spiral spring is stored up in the spring, and when the spring is allowed to contract to its original length it does an equal amount of work.

It is found that energy can exist in many different forms, e.g. heat, electrical energy, magnetic energy and chemical energy. It can be made to change from one form to another, but it is never destroyed or created.

Experimental evidence for the truth of the principle of the conservation of energy will be described in later chapters dealing with different forms of energy and its transformations from one form to another.

Energy which depends on the arrangement of the parts of a system is called potential energy. For example, the energy possessed by a spiral spring when it has been stretched is potential energy, for it depends on the arrangement of the parts of the spring. A system consisting of a heavy body and the earth possesses potential energy, for the heavy body can do work when it is allowed to move down nearer to the earth. When a force f acts on a body of mass m which is free to move, the force imparts to it an acceleration a given by the equation $f = ma$. Suppose the body moves along a straight line in the direction of the force. Then the distance s described by the body in a time t is given by the equation

$$s = \tfrac{1}{2}\,(v_2 + v_1)\,t,$$

where v_1 denotes the initial and v_2 the final velocity of the body. We have also

$$f = m\,\frac{v_2 - v_1}{t}.$$

The work w done by the force on the body is given by the equation

$$fs = \tfrac{1}{2}\,(v_2 + v_1)\,tm\,\frac{v_2 - v_1}{t},$$

or $$w = \tfrac{1}{2}m\,(v_2{}^2 - v_1{}^2).$$

Thus the work done is equal to the increase in $\tfrac{1}{2}mv^2$, where m is the mass and v the velocity of the body. This quantity is called the kinetic energy of the body. It is the energy which the body possesses in virtue of its motion and is equal to the work which the body can do before being brought to rest. For suppose that a body of mass m moving along a straight line with velocity v is acted on by a force in the opposite direction to that in which it is moving. After a time t its velocity will be reduced to zero. We have $s = \frac{1}{2}vt$ so that the work done in the time t is given by

$$w = fs = m\,\frac{v}{t} \times \frac{v}{2}\,t = \frac{1}{2}\,mv^2.$$

So far as we know all the different forms of energy such as heat and electrical energy can be regarded as partly potential and partly kinetic. The energy of a system of material particles in motion is partly potential and partly kinetic. The potential energy at any instant is the same as it would be if the particles were all at rest in their positions at the instant considered and the kinetic energy is equal to

$$\tfrac{1}{2}m_1v_1^2 + \tfrac{1}{2}m_2v_2^2 + \tfrac{1}{2}m_3v_3^2 + \ldots,$$

where m_1, m_2, m_3, etc. denote the masses of the particles and v_1, v_2, v_3, etc. their velocities. If P denotes the potential energy of a system and K its kinetic energy then, so long as no external actions take place, we have $P + K = \text{constant}$. This equation is the mathematical expression of the principle of the conservation of energy.

When work is being done the rate of doing it or the work done per unit time is called the *power*. If an

Power.

amount of work w is done in a time t at a uniform rate then $p = \dfrac{w}{t}$, where p denotes the power.

The unit of power is one unit of work done in unit time. The unit of power may be denoted by ML^2T^{-3}, where M, L and T as usual denote the units of mass, length and time.

The unit of power most frequently used by engineers is called one horse-power and is equal to 33,000 foot-pounds of work per minute or 550 foot-pounds per second. Another unit of power used in electrical engineering is called one Watt and is equal to ten million ergs per second. One horse-power is equal to nearly 746 Watts. A kilowatt is one thousand Watts, and is equal to 1·34 horse-power.

If the point at which a force f acts is moving with a velocity v in the direction of the force, the work done by the force in unit time is fv, which is therefore equal to the power developed by the action of the force. Consider the case of water flowing along a pipe. Suppose the water is under a pressure P, that is, let P denote the force with which the water presses *on each unit area* with which it is in contact. Consider a cross section AB (Fig. 31) of the pipe. The water to the left of AB is exerting a force on the water to the right of AB which is equal to Pa, if a denotes the area of the cross section AB. Suppose that at AB the water is moving from left to right with a velocity v. Then the work done by the water to the left of AB on that to the right in unit time is equal to Pav, because v is the distance the water moves in unit time. The power represented by the stream of

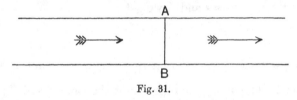

A

B

Fig. 31.

water, that is, the work done per second at the cross section AB, is therefore equal to Pav. But av is the volume V of water flowing past AB in unit time. The power of a stream of water is therefore equal to PV, or its pressure multiplied by the volume flowing in unit time. For example, the power of 1000 cubic feet of water per minute at a pressure of 330 pounds weight on the square foot is equal to 330,000 foot-pounds per minute or to ten horse-power. In this calculation of the power we have neglected the kinetic energy of the stream of water.

There are a number of mechanical appliances usually referred to as machines in which a force acting at one point causes the machine to exert a greater or less force at another point. If the machine moves, then the force acting on the machine and the force exerted by the machine both do work. Let f denote the force acting on the machine, s the distance through which it acts, f' the force exerted by the machine and s' the distance through which it acts. Then the work done

Machines.

on the machine is fs and the work done by the machine is $f's'$. If no energy is lost or stored up in the machine, it follows from the principle of the conservation of energy that

$$fs = f's'.$$

The efficiency of a machine is the ratio of $f's'$ to fs. If $fs = f's'$ the efficiency is equal to unity and the machine is said to be perfectly efficient. In practice some of the work done on the machine is wasted in it in overcoming frictional forces, so that $f's'$ is always less than fs. The efficiency is therefore always less than unity. The equation $fs = f's'$ can be used to calculate the theoretical value of f', which would be obtained if the machine were perfectly efficient. The actual value must be less than that so calculated.

If we divide the equation $fs = f's'$ by t, the time taken to move through the distances s and s', we get

$$f\frac{s}{t} = f'\frac{s'}{t}.$$

Let $\dfrac{s}{t} = v$ and $\dfrac{s'}{t} = v'$, so that $fv = f'v'$. Here v and v' are the velocities of the points of application of the forces f and f'; so we see that, for a perfectly efficient machine, the power acting on the machine is equal to the power the machine exerts. The machine therefore serves to transmit power from one point to another. The efficiency of the machine is equal to the power it gives out divided by the power it receives, for

$$\frac{f's'}{fs} = \frac{f'v'}{fv}.$$

The simplest type of machine is the lever. This consists of a rigid bar which is held fixed at one point about which it can turn. Suppose QP (Fig. 32) is a lever which can turn about F. Let

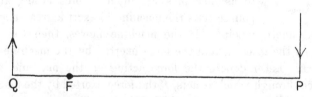

Fig. 32.

a force f act at P in a direction perpendicular to the bar and let
the bar exert a force f' at Q also perpendicular to the bar. Now
let the bar turn through a very small angle θ. P will move
through a distance $FP \times \theta$, so that the work done by f will be
$FP \times \theta f$. In the same way the work done at Q will be $FQ \times \theta f'$.
Hence we have

$$FP \times \theta f = FQ \times \theta f', \text{ or } f' = f \frac{FP}{FQ}.$$

f and f' may both be on the same side of F instead of on opposite
sides.

Another simple machine is the pulley, which consists of a wheel
over which a flexible cord or belt is passed. Let P (Fig. 33) be a

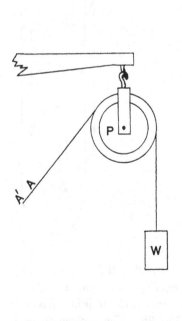

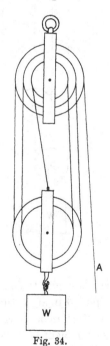

Fig. 33.   Fig. 34.

pulley, and suppose a weight W is hung from one end of the cord
and that this weight is raised by pulling the other end A. If A
is moved from A to A' the weight will be raised through a
height equal to AA'. The force required at A to balance the

weight is therefore equal to the weight, provided there is no friction in the pulley. The tension in the cord is therefore the same on both sides of the pulley.

Various combinations of pulleys are employed in practice. Consider for example the combination shown in Fig. 34. If the string at A is pulled down with a force T there will be a tension equal to T throughout the whole length of the string, so that since the weight W is supported by five strings the upward force on W

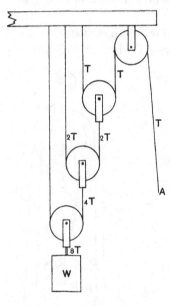

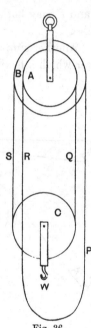

Fig. 35.   Fig. 36.

will be equal to $5T$. Also it is easy to see that if five feet of string is pulled in at A, the weight W will only be raised one foot, so that the work done at A is equal to that done on the weight. Another arrangement of pulleys is shown in Fig. 35. The tension in each string is marked in the figure; so we see that a tension T at A supports a weight $W = 8T$. What is called a differential pulley is shown in Fig. 36. Two pulleys B and A, one slightly larger than the other, are fixed together on the same axle. A chain passes over B round the lower pulley C and then over A. The pulleys

A and B are made with cogs which fit into the links of the chain so that it cannot slip on them. If the chain at P is pulled down through a distance d then the chain at S is pulled up an equal distance, but the chain at Q is let down by the rotation of A and B through a distance $d\dfrac{r_A}{r_B}$, where r_A denotes the radius of A and r_B that of B. Consequently C is raised by an amount $\frac{1}{2}\left(d - d\dfrac{r_A}{r_B}\right)$.

The force f at P therefore gives a force at W equal to $2f\dfrac{r_B}{r_B - r_A}$. For example if $r_B = 10$ inches and $r_A = 9{\cdot}5$ inches then the force at W is 40 times that at P, but P has to be pulled down 40 times as far as W moves up.

A machine often used in practice is the screw. It consists of a circular cylinder on the surface of which a spiral groove is cut.

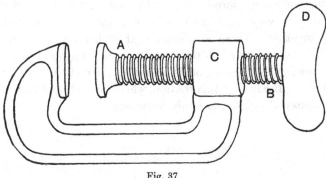

Fig. 37.

The ridge between the adjacent parts of the groove is called the thread of the screw. The distance measured parallel to the axis of the cylinder from one thread to the next is called the pitch of the screw. The pitch should be constant throughout the length of the screw. Screws are made by rotating the cylinder while a cutting tool moves along parallel to the axis of the cylinder. The tool is made to move through a distance proportional to the angle through which the cylinder turns. The screw is used in combination with a nut, which is a plate with a hole in it which is grooved so as to fit the screw.

Fig. 37 shows a screw clamp which is used for pressing bodies firmly together. A screw AB turns in a nut at C. If the nut is held fixed and the screw turned round once, the screw advances through the nut a distance equal to its pitch. If the screw is turned by applying a force f at D, perpendicular to the plane containing D and the axis of the screw, then if the screw makes one revolution the force f acts through a distance $2\pi r$, where r is the distance of D from the axis. The end of the screw at A then advances a distance p equal to the pitch. The force f' exerted by the screw at A is therefore given by the equation

$$2\pi r f = p f', \text{ or } f' = f \frac{2\pi r}{p}.$$

For example, if $r = 2$ feet and $p = \frac{1}{4}$ inch we get theoretically, assuming no friction,

$$f' = f \times 2\pi 2 \times 48 = 604 f.$$

Thus very large forces can be obtained with screws. Screws and nuts are very much used for holding the parts of machinery firmly together. There is a great deal of friction between the screw and its nut, so much that usually a force applied to the end of a screw in the direction of its axis will not make it turn round in its nut. Because of this friction, when the nut on a screw has been tightened it does not easily work loose.

CHAPTER VI

MECHANICS OF RIGID BODIES

IF a rigid body such as a wheel is mounted on a cylindrical
Moment of a shaft which can turn round in fixed bearings, then
Force. the body is free to rotate about the axis of the shaft
but cannot move in any other way. In this chapter we shall
consider chiefly the motion of such a body one straight line in
which is fixed so that the body can only rotate round this line as
axis. Let AB (Fig. 38) be a rigid body, and let the axis round

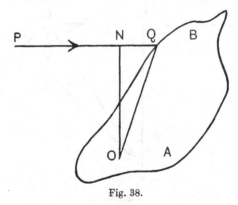

Fig. 38.

which it is free to turn be perpendicular to the plane of the paper
and pass through O. Let a force act on the body at the point Q.
We can resolve the force into two components, one parallel to the
axis through O and the other in the plane of the paper. The
component parallel to the axis can do no work on the body because
the body when it rotates does not move in the direction of this
component, which therefore has no effect on the motion of the
body and so need not be further considered. Let PQ represent
the component in the plane of the paper and let its magnitude be
denoted by f. From O draw ON perpendicular to PQ.

Suppose now that the body turns round through a very small angle θ so that Q moves to Q' (Fig. 39). Let $OQ = r$ and $ON = p$. Then $QQ' = r\theta$.

The work done by the force f is equal to f multiplied by the resolved part of QQ' in the direction parallel to PQ. Draw $Q'M$ parallel to QP and drop QM perpendicular to $Q'M$. The work done by f is then equal to $f \times MQ'$. Since θ is very small we have $N\hat{Q}O = M\hat{Q}Q'$ so that the triangles MQQ' and NQO are similar. Hence

$$MQ' = QQ' \times \frac{NO}{OQ} = QQ' \frac{p}{r}.$$

The work done by f is therefore equal to $fQQ'\frac{p}{r}$; but $QQ' = r\theta$, so the work is equal to $fp\theta$. $p\theta$ is equal to the distance N would

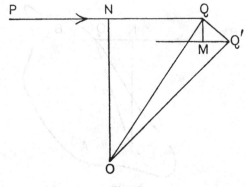

Fig. 39.

move parallel to PQ if ON turned through the very small angle θ. Thus the work done by the force is proportional to p the perpendicular distance from the axis of rotation on to the line of action of the force. If this line of action passes through the axis, $p = 0$ and the force can do no work on the body and so cannot affect its motion. The product fp is called the moment of the force f about the axis through O.

Suppose that another force acts on the body at R. As before we need only consider the component of this force in a plane perpendicular to the axis, for the component parallel to the axis can have no effect on the motion of the body. Let SR (Fig. 40) represent the component parallel to the plane of the paper and let it be equal to f'. Let p' be the length of the perpendicular between SR and the axis. If the body turns through a small angle θ, the work done by f' is $f'p'\theta$. If the total work $fp\theta + f'p'\theta$ is equal to zero, the effect of the force f' balances the effect of the force f.

For if no work is done on the body its kinetic energy remains unchanged so that it moves with a constant velocity, which shows that the resultant action on it is zero. If therefore $fp = -f'p'$ the two forces f and f' balance each other, so that the body if at rest would be in equilibrium. We see therefore that the moment of a force about an axis is the proper measure of the action of the force in tending to turn a body round the axis. In calculating the moment of any force about an axis we first find its component in a plane perpendicular to the axis and then multiply this component by the length of the perpendicular in this plane from the axis on to

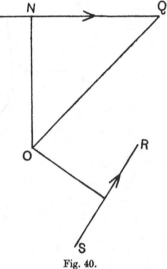

Fig. 40.

the component in question. If any number of forces act on the body, then the condition of equilibrium is that the sum of all their moments about the axis shall be zero. The moments of forces which turn the body one way are counted positive and the moments of the others which turn it the opposite way are counted negative.

Suppose that two parallel forces f and f' act on the body, both in a plane perpendicular to the axis about which it is free to rotate.

Parallel Forces.

Let PQ (Fig. 41) and RS represent the forces f and f', and let p and p' be the lengths of the perpendiculars from O on to PQ and RS. If $fp + f'p' = 0$ the forces do not tend to turn the body round the axis at O. These forces however will tend to move the

W. P.   5

body parallel to their own directions.   This motion is stopped by
the axis which we are supposing is fixed.   The force on the axis,

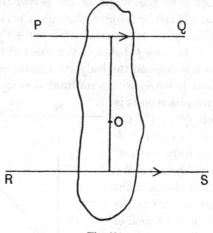

Fig. 41.

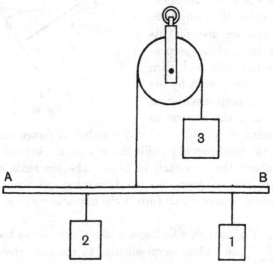

Fig. 42.

which it resists with an equal and opposite reaction, is evidently
equal to $f+f'$.   This force $f+f'$ acting through O is the resultant
of the two parallel forces f and f'; so that we see that the resultant

of two parallel forces is equal to their sum and acts along a line in their plane such that their total moment about any axis through this line is zero.

For example, let a light rod AB (Fig. 42) be hung up by a string at its middle point and two weights of, say, two pounds and one pound be hung from the rod. Then to keep the rod balanced, the one pound weight must be hung twice as far from the centre of the rod as the two pound weight; and the tension in the string supporting the rod will be equal to three pounds weight.

Suppose now that two equal but oppositely directed parallel forces act on a body which is free to turn about **Couples.** an axis at O (Fig. 43). Let the plane containing the two forces be perpendicular to the axis. Let PQ

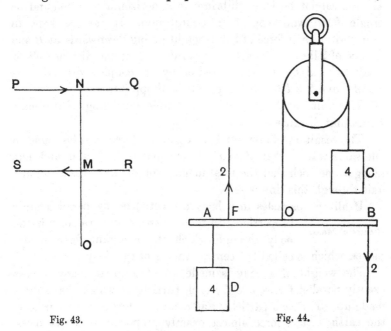

Fig. 43.   Fig. 44.

represent one of the forces f and let RS represent the other $-f$. Let $ON = p$, $OM = p'$, and let the distance between the forces $MN = d$. The moment of f about O is fp and that of $-f$ is $-fp'$. The total moment is therefore $fp - fp' = f(p - p') = fd$.

5—2

The total moment fd is therefore the same for any position of
the axis O; so that the total moment of two equal and opposite
parallel forces is the same about any axis perpendicular to their
plane and is equal to either force multiplied by the distance
between them. Their resultant is equal to $f-f$ and so is zero.
Such a pair of forces is called a couple. The product fd is called the
moment of the couple, or sometimes it is called simply the couple.

A couple tends to produce rotation, but since its resultant is
zero it does not tend to produce any translation. A couple can
only be balanced by an equal and opposite couple, it cannot be
balanced by a single force.

For example, if a light rod AB (Fig. 44) is hung up at O by a
string passing over a pulley as shown, to which a 4 lb. weight C is
attached, and if another 4 lb. weight D is hung from the rod, then
the rod cannot be in equilibrium in a horizontal position and no
single force can keep it in equilibrium. It can be kept in
equilibrium by a force of 2 lbs. weight acting downwards at B and
a force of 2 lbs. weight acting upwards at F if the distance FB is
made twice AO. It is then acted on by two couples with moments
$4 \times AO$ and $2 \times FB$ which are equal and opposite.

The unit couple is equal to a couple consisting of two unit
forces unit distance apart.

The resultant of any number of parallel forces can be found in
the same way as that of two. It is equal to their sum and acts
along a line such that the total moment of all the forces about any
axis through this line is zero.

If all the particles in a body are acted on by parallel forces
proportional to their masses, there is a point in the
Centre of Mass. body through which the resultant force always
passes, which is called the centre of mass of the body.

The weight of a body is made up of a great many almost
exactly parallel forces, one for each particle of which the body is
made up. For each particle is attracted vertically downwards by
the earth with a force almost exactly proportional to its mass.
The point through which the resultant of the weights of all the
particles acts is called the centre of gravity of the body. Except
for exceedingly large bodies the centre of gravity and the centre of
mass practically coincide.

Let us first consider the position of the centre of mass of two particles A and B of masses m_1 and m_2 at a distance d apart (Fig. 45). Let the forces acting on them be equal to αm_1 and αm_2 and act along AA' and BB'. The resultant of these two forces acts along a line CC' in the same plane as AA' and BB' and parallel to them. Let it cut BA at C. Draw BMN perpendicular to CC' and AA'. Then we have $BM \times \alpha m_2 = NM \times \alpha m_1$. But

$$\frac{BC}{CA} = \frac{BM}{MN},$$

so that $$BC \times \alpha m_2 = CA \times \alpha m_1.$$

The resultant therefore cuts the line AB at a point C such that

$$\frac{BC}{CA} = \frac{m_1}{m_2}.$$

The position of C is independent of α and of the direction in which the forces αm_1 and αm_2 act. C is therefore the centre of mass of A and B.

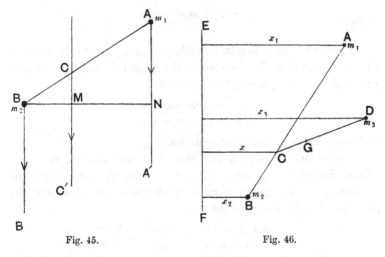

Fig. 45.   Fig. 46.

Take any plane surface EF (Fig. 46) near AB and draw perpendiculars to it from A, B and C. Let their lengths be x_1, x_2 and x respectively. We have

$$\frac{BC}{CA} = \frac{m_1}{m_2} = \frac{x - x_2}{x_1 - x}.$$

Hence $\qquad m_1 x_1 - m_1 x = m_2 x - m_2 x_2 ,$

or $\qquad x = \dfrac{m_1 x_1 + m_2 x_2}{m_1 + m_2} .$

Suppose now we have a third particle of mass m_3 at D. The resultant of the three forces αm_1, αm_2 and αm_3 can be got by finding the resultant of αm_3 and the resultant of αm_1 and αm_2 which acts at C and is equal to $\alpha (m_1 + m_2)$. The resultant of the three forces will therefore always pass through a point G on CD such that

$$\frac{GD}{CG} = \frac{m_1 + m_2}{m_3} .$$

This point G is the centre of mass of m_1, m_2 and m_3. Let the perpendicular distance from G on to the plane EF be x' and that of D be x_3. Then as before

$$x' = \frac{(m_1 + m_2) x + m_3 x_3}{(m_1 + m_2) + m_3} .$$

But $\qquad (m_1 + m_2) x = m_1 x_1 + m_2 x_2 ,$

so that $\qquad x' = \dfrac{m_1 x_1 + m_2 x_2 + m_3 x_3}{m_1 + m_2 + m_3} .$

If now we take a fourth particle of mass m_4 we can show in the same way that the distance of the centre of mass of the four particles from the plane EF is equal to

$$\frac{m_1 x_1 + m_2 x_2 + m_3 x_3 + m_4 x_4}{m_1 + m_2 + m_3 + m_4} ,$$

and so on for any number of particles.

If therefore we have n particles of masses $m_1, m_2, m_3, \ldots m_n$ whose distances from any plane are $x_1, x_2, x_3, \ldots x_n$, then the distance x of the centre of mass of all these particles from the plane is given by the equation

$$x = \frac{m_1 x_1 + m_2 x_2 + \ldots + m_n x_n}{m_1 + m_2 + \ldots + m_n} = \frac{\Sigma m x}{\Sigma m} .$$

The centre of mass of two equal masses lies half-way between them ; so that if a body is symmetrical about a plane its centre of mass will be in that plane. Thus the centre of mass of a uniform rod is at the middle point of the rod and the centre of mass of a solid sphere of uniform density is at the centre of the sphere.

If a body is hung up by a thread, then it is kept in equilibrium by the tension in the thread acting upwards and its weight acting downwards. The resultant of the weights of all the particles making up the body must therefore be equal and opposite to the tension in the thread. The direction of the thread will therefore pass through the centre of gravity of the body, which practically coincides with its centre of mass. If the direction of the thread is marked on the body and it is then attached by a different point to the thread and hung up again, the new direction of the thread in it will cut the old direction at the centre of gravity. This gives an experimental method of finding the position of the centre of gravity or the centre of mass of a body.

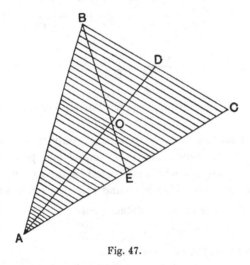

Fig. 47.

Let ABC (Fig. 47) be a uniform triangular sheet and suppose we require to find the position of its centre of mass. Draw a large number of equidistant lines on it all parallel to BC. Bisect BC at D and join AD. The centre of mass of each of the narrow strips parallel to BC lies on AD; so that the centre of mass of the whole plate must lie on AD. Now bisect AC at E and join BE. Let AD and BE cross at O. In the same way as for AD we see that the centre of mass must lie on BE. Therefore the centre of mass of the triangle must be at O. The point O is twice as far from A as it is from D.

This result can also be obtained by means of the formula $x = \dfrac{\Sigma mx}{\Sigma m}$. Take a plane passing through A and perpendicular to AD. If we suppose the mass of each of the strips parallel to BC to be concentrated at their middle points, we have at a series of equidistant points along AD a series of masses proportional to their distances from A.

Let $AD = l$, and let there be n strips. The distances of the masses from A are therefore $l/n,\ 2l/n,\ 3l/n,\ \dots\ nl/n$. Let the masses of the strips be $m,\ 2m,\ 3m,\ \dots\ nm$. The distance of the centre of mass from A is therefore given by the equation

$$\frac{\Sigma mx}{\Sigma m} = \frac{\dfrac{lm}{n}(1^2 + 2^2 + 3^2 + \dots + n^2)}{m(1 + 2 + 3 + \dots + n)}$$

$$= \frac{\dfrac{l}{n}\, m\, \dfrac{n^3}{3}}{m\, \dfrac{n^2}{2}} = \frac{2}{3}\, l,$$

when n is made indefinitely large *. This agrees with the previous result.

When a force acts on a body which is only free to turn about an axis there is an equal and parallel force on the axis which is resisted by the axis with an equal and opposite reaction. The force and this reaction form a couple.

Reaction of Axis.

* To show that when n is made indefinitely large

$$1^m + 2^m + 3^m + \dots + n^m = \frac{n^{m+1}}{m+1},$$

let the series be noted by S_n and assume that

$$S_n = A_0 + A_1 n + A_2 n^2 + A_3 n^3 + \text{etc.} \dots\dots\dots\dots\dots\dots(1).$$

Change n to $n + 1$ so that

$$S_{n+1} = S_n + (n+1)^m = A_0 + A_1(n+1) + A_2(n+1)^2 + \text{etc.}\dots\dots\dots(2).$$

Subtracting (1) from (2) and expanding $(n+1)^m$, we get

$$n^m + mn^{m-1} + \frac{m(m+1)}{1.2}n^{m-2} + \dots = A_1 + A_2(2n+1) + A_3(3n^2 + 3n + 1) + \text{etc.}$$

This equation is true for all values of n, so that the coefficients of each power of n on both sides of it must be equal. The highest power of n on the left-hand side is n^m, so that all the A's of higher order than A_{m+1} are zero. If n is made indefinitely large we can neglect all powers of n except the highest and the equation reduces to $A_{m+1}(m+1)n^m = n^m$.

Hence $S_n = \dfrac{n^{m+1}}{m+1}$ when n is made indefinitely large.

The reaction of the axis does not tend to turn the body round because it has no moment about the axis. The moment of the force acting on the body about the axis is therefore equal to the moment of the couple formed by this force and the reaction of the axis.

A constant angular velocity is equal to the angle turned through in unit time. It is usually denoted by ω, so that $\theta = \omega t$ where θ denotes the angle turned through in a time t. If the angular velocity is changing at a uniform rate, then the change in unit time is called angular acceleration. If α denotes the angular acceleration

$$\alpha = \frac{\omega_2 - \omega_1}{t},$$

where ω_1 is the angular velocity at the beginning and ω_2 that at the end of a time t. The angle θ described in a time t is given by the equation

$$\theta = \frac{\omega_2 + \omega_1}{2} t = \omega_1 t + \tfrac{1}{2}\alpha t^2.$$

Also

$$\theta\alpha = \frac{\omega_2 + \omega_1}{2} t \; \frac{\omega_2 - \omega_1}{t} = \tfrac{1}{2}(\omega_2^2 - \omega_1^2).$$

In dealing with the rotation of bodies it is usually more convenient to measure angles in circular measure than in degrees.

Rotation of a Rigid Body. When a rigid body is rotating with an angular velocity ω about a fixed axis, then the velocity of a point P in the body at a distance r from the axis is equal to $r\omega$. If the body has an angular acceleration α, then the point P has an acceleration equal to $r\alpha$ along the direction in which it is moving, that is, along the tangent at P to the circle which P is describing about the axis of rotation. (It has also the acceleration v^2/r along the radius vector.)

A particle of mass m at P must therefore be acted on by a force having a component equal to $mr\alpha$ along the tangent at P. The moment of this force about the axis is $mr\alpha \times r = mr^2\alpha$.

Let us now regard the rigid body as divided by imaginary surfaces into an immense number of very small parts. Let the masses of these parts be denoted by m_1, m_2, m_3, etc., and let their distances from the axis be denoted by r_1, r_2, r_3, etc. The total mass M of the body is then given by

$$M = m_1 + m_2 + m_3 + \text{etc.} = \Sigma m,$$

where Σm denotes the sum of all the masses m_1, m_2, m_3, etc. The nth particle requires a force $m_n r_n \alpha$ to give it its acceleration $r_n \alpha$ and the moment of this force about the axis is $m_n r_n^2 \alpha$. The total moment C required to give the whole body the angular acceleration α is therefore given by

$$C = m_1 r_1^2 \alpha + m_2 r_2^2 \alpha + m_3 r_3^2 \alpha + \text{etc.} = \Sigma m r^2 \alpha,$$

or $\qquad C = \alpha \Sigma m r^2.$

The forces acting on the particles of the body are either forces due to actions between one particle and another or external forces acting on the body from outside. Any internal action is made up of two equal and opposite forces the moments of which about the axis are therefore equal and opposite. The total moment of the internal forces about the axis is therefore zero; so that the total moment of the forces acting on all the particles of the body is equal to the moment about the axis of the external forces acting on the body.

If then C denotes the total moment of the external forces acting on the body about its axis of rotation, we have $C = \alpha \Sigma m r^2$.

The quantity $\Sigma m r^2$ is called the *moment of inertia* of the body
Moment of Inertia.   about the axis in question. The moment of inertia of a rigid body about any axis is obtained by supposing the body divided into a great many very small parts and taking the sum of the products formed by multiplying the mass of each part by the square of its distance from the axis.

It should be remembered that the moment of inertia of a body about an axis depends on the position of the axis with respect to the body, so that in stating the moment of inertia of a body the position of the axis in question must be specified.

If we suppose the body divided up into a very large number n of particles all having equal masses m, then the moment of inertia

$$\Sigma m r^2 = m \Sigma r^2,$$

where $\qquad \Sigma r^2 = r_1^2 + r_2^2 + r_3^2 + \ldots + r_n^2.$

Let the average value of r^2 be equal to k^2 so that $\Sigma r^2 = n k^2$.

The moment of inertia is therefore equal to mnk^2 or Mk^2, if M denotes the total mass of the body. The moment of inertia is therefore the same as that of a particle of mass equal to that of the body at a distance k from the axis. k is called the radius of gyration of the body about the axis in question. k is given by the equation

$$\Sigma mr^2 = k^2 \Sigma m$$

or

$$k^2 = \frac{K}{M},$$

where K denotes the moment of inertia of any rigid body and M its mass. The equation $C = a\Sigma mr^2$ may be written $C = K\alpha$, or since

$$\alpha = \frac{\omega_2 - \omega_1}{t},$$

$$C = \frac{K\omega_2 - K\omega_1}{t}.$$

As an example consider the case of a simple pendulum consisting of a particle P (Fig. 48) of mass m suspended from a fixed point O by a light thread of length l. Let OC be a vertical line and draw PN perpendicular to OC. If the pendulum is swinging through a small angle in the plane of the paper the thread remains straight, so we may regard the pendulum as a rigid body. The forces acting on P are the tension in the thread, which has no moment about the axis through O about which P rotates, and the weight of P. The moment of the weight about an axis through O perpendicular to the plane in which P is swinging is $mg \times PN$. The moment of inertia of P about this axis is ml^2. Hence we have, since $C = a\Sigma mr^2$,

$$- mg . PN = - mgl \sin \theta = \alpha ml^2,$$

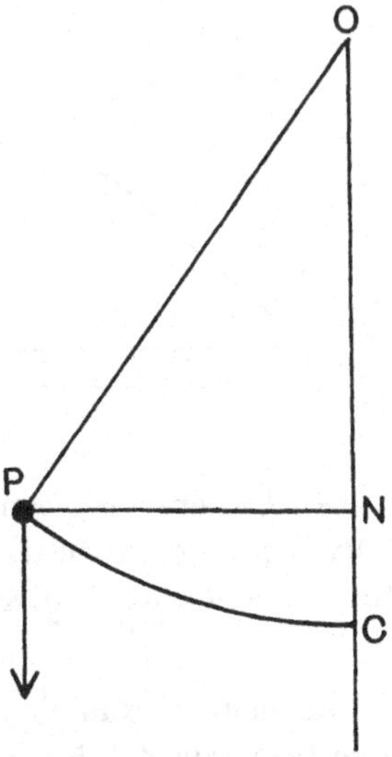

where θ is the angle PON and α the angular acceleration of OP round the axis.

If θ is small we may put $\theta = \sin \theta$, and so get $- g\theta = \alpha l$.

Now the acceleration a of P along its direction of motion is

Fig. 48.

equal to αl and its distance x from C along the arc CP is equal to θl, so that

$$-g\,\frac{x}{l} = a.$$

This is the same result as that obtained in Chapter IV, and from this it follows as before that the time T of a complete vibration is given by

$$T = 2\pi\sqrt{\frac{l}{g}}.$$

If we put $\dfrac{g}{l} = c$, we get $\alpha = -c\theta$ and $T = 2\pi/\sqrt{c}$.

We see from this that if a body is rotating about an axis in such a way that its angular acceleration α is given by the equation

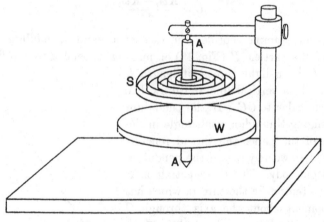

Fig. 49.

$\alpha = -c\theta$, where c is a constant and θ its angular displacement from a fixed position, then it oscillates about the fixed position and its time of oscillation is given by

$$T = 2\pi/\sqrt{c}.$$

As another example consider a wheel W (Fig. 49), mounted on a vertical axle AA, having fastened to it one end of a spiral spring S, the other end of the spring being fixed. If the wheel is turned round, the spring exerts a force on it which tends to bring it back to its original position of equilibrium. The moment of this force about the axis of the wheel is proportional to the angle through which the wheel is turned. Let this moment then be equal to $-A\theta$, where A is a constant. Let the moment of inertia of the wheel about its axis of rotation be $K = \Sigma mr^2$. Then we have

$$-A\theta = \alpha K,$$

where α is the angular acceleration of the wheel. If the wheel is turned round and then let go, it will therefore oscillate about its position of equilibrium, and the time of a complete vibration will be

$$T = 2\pi\sqrt{\frac{K}{A}}.$$

The balance wheel of a watch is an example of such a wheel vibrating under the action of a spiral spring. The time of vibration is the same for small vibrations as for large ones and so the watch goes at the same rate whatever the amplitude of vibration of its balance wheel.

If a small body of mass m is attached to the wheel just considered, at a distance r from the axis, then the moment of inertia will be increased from K to $K + mr^2$. The time of oscillation will therefore be increased to

$$T' = \sqrt{\frac{K + mr^2}{A}}.$$

If we measure T, T', m and r^2, we can find K and A from the last two equations. We get by dividing the two sides of one equation by the corresponding sides of the other and squaring

$$\left(\frac{T'}{T}\right)^2 = \frac{K + mr^2}{K},$$

or

$$K = \frac{mr^2}{\left(\dfrac{T'}{T}\right)^2 - 1}.$$

The body of mass m may be in the form of a thin hollow circular cylinder of radius r, attached to the wheel so that its axis coincides with the axis of rotation. All the matter in the cylinder is then at the same distance r from the axis, and so its moment of inertia about the axis is mr^2. This gives an accurate way of finding experimentally the moment of inertia of a body about an axis. Instead of mounting the body on an axle and using a spiral spring it may be simply hung up by a wire. When the body is turned round the wire is twisted and resists the motion with a couple proportional to the angle of the twist.

The moments of inertia of bodies having simple geometrical forms can be calculated by means of the formula $K = \Sigma mr^2$. As an example take the case of a thin uniform straight rod of length l and mass m. Let the axis about which the moment of inertia is required be perpendicular to the rod and pass through its middle point. Suppose the rod divided into a very large number n of equal parts each of length l/n and mass m/n. The moment of inertia is then equal to

$$2\frac{m}{n}\left(\frac{l}{n}\right)^2\left\{1^2 + 2^2 + 3^2 + \ldots + \left(\frac{n}{2}\right)^2\right\} = 2\frac{m}{n}\left(\frac{l}{n}\right)^2\frac{n^3}{24} = \frac{ml^2}{12},$$

for
$$1^p + 2^p + 3^p + \ldots + q^p = \frac{q^{p+1}}{p+1}$$

when q is a very large number.

Now consider the moment of inertia of a circular disk of mass m and radius a about an axis through its centre and perpendicular to its plane. Its mass per unit area is $m/\pi a^2$. Suppose it is divided up into circular rings of radii a/n, $2a/n$, $3a/n$ and so on where n is a very large number. The area of the pth ring is equal to

$$2\pi\frac{pa}{n}\times\frac{a}{n}.$$

The moment of inertia of any ring about the axis in question is equal to the mass per unit area multiplied by the area of the ring and multiplied by the square of the radius of the ring. The moment of inertia of the disk is therefore equal to

$$\frac{m}{\pi a^2}2\pi\frac{a^2}{n^2}\frac{a^2}{n^2} + \frac{m}{\pi a^2}2\pi\frac{2a^2}{n^2}\frac{4a^2}{n^2} + \ldots + \frac{m}{\pi a^2}2\pi\frac{pa^2}{n^2}\frac{p^2a^2}{n^2} + \ldots$$

$$= \frac{2ma^2}{n^4}\{1^3 + 2^3 + \ldots + n^3\} = \frac{2ma^2}{n^4}\frac{n^4}{4} = \frac{ma^2}{2}.$$

The moment of inertia of a thin circular disk about one of its diameters is equal to $ma^2/4$. Moments of inertia are most easily calculated by the mathematical process known as integration.

If the moment of inertia of a body of mass M, about an axis which passes through its centre of mass, is K, then its moment of inertia about a parallel axis, at a distance d from the first axis, is equal to $K + Md^2$.

Let the two axes be perpendicular to the plane of the paper and cut it at O and P (Fig. 50). Consider one of the particles which make up the body. Let its mass be m and let it be at A in the plane of the paper. Let $OA = r$. The moment of inertia of this particle about the axis at O is mr^2 and about the axis at P it is $m(AP)^2$. Draw AN perpendicular to OP and let $ON = x$. We have

$$AP^2 = AN^2 + NP^2 = AN^2 + (OP - ON)^2$$
$$= AN^2 + ON^2 + OP^2 - 2OP \cdot ON$$
$$= r^2 + d^2 - 2dx.$$

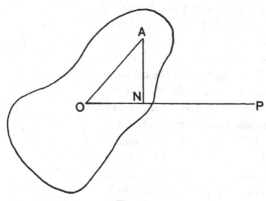

Fig. 50.

Take a plane through O perpendicular to OP. Let the body be made up of particles of masses m_1, m_2, m_3, ... m_n. Let their distances from the plane through O be x_1, x_2, x_3, ... x_n and their distances from the axis at O be r_1, r_2, r_3, ... r_n. Then the moment of inertia about the axis at O is $\Sigma mr^2 = K$. For any particle, say the mth, we have, as for the one first considered, the square of its distance from the axis at P equal to $r_m^2 + d^2 - 2dx_m$. The moment of inertia about the axis at P is therefore equal to

$$\Sigma m (r^2 + d^2 - 2dx),$$
$$= \Sigma mr^2 + d^2 \Sigma m - 2d\Sigma mx,$$
$$= K + Md^2 - 2d\Sigma mx.$$

The distance of the centre of mass of the body from the plane through O perpendicular to OP is $\dfrac{\Sigma mx}{\Sigma m}$. But the centre of mass is at O so that $\dfrac{\Sigma mx}{\Sigma m} = 0$ and therefore $\Sigma mx = 0$. Hence the moment of inertia about the axis through P is equal to $K + Md^2$.

If a rigid body is free to turn only about a horizontal axis and Compound Pendulum. is acted on by no external forces except its weight and the reaction of the axis, then it can only rest in equilibrium if its centre of gravity is vertically below the axis. For otherwise the weight has a moment about the axis which imparts to the body an angular acceleration. Let the horizontal axis be perpendicular to the plane of the paper and pass through O (Fig. 51). Draw OA vertically downwards. Let G be the centre of gravity of the body and let $OG = h$. The resultant of the weights of all the particles which make up the body acts downwards through G and is equal to Mg, where M is the mass of the body. Let the angle $GOA = \theta$. The moment of the weight about the axis at O is equal to $- Mgh \sin \theta$. Let the moment of inertia of the body about the axis at O be K. Then we have

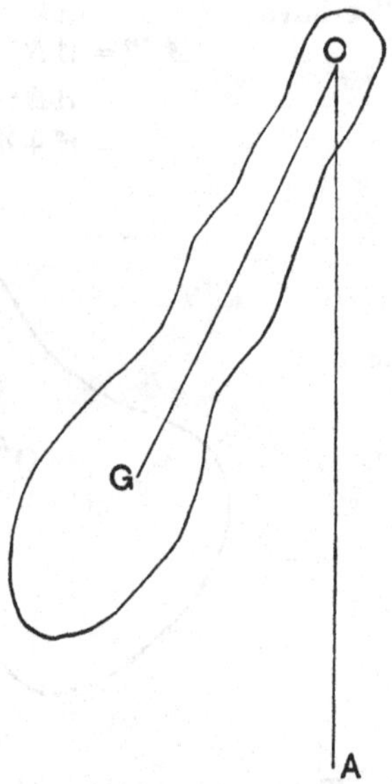

$$- Mgh \sin \theta = K\alpha,$$

where α denotes the angular acceleration of the body. If the angle θ is very small we may put $\sin \theta = \theta$ and so get $\alpha = \dfrac{Mgh\theta}{K}$. It appears therefore that the body has an angular acceleration proportional to its angular displacement and in the opposite direction to the displacement. The body therefore moves with a simple harmonic motion and the time of a complete vibration is given by

$$T = 2\pi \sqrt{\frac{K}{A}} = 2\pi \sqrt{\frac{K}{Mgh}}.$$

Fig. 51.

A rigid body oscillating through a small angle about a horizontal axis under the action of its weight is called a compound pendulum.

Let the moment of inertia of the body about an axis through G parallel to the axis at O be equal to Mk^2, where k is the radius of gyration about the axis at G. Then we have

$$K = Mk^2 + Mh^2.$$

Hence
$$T = 2\pi \sqrt{\frac{h^2 + k^2}{gh}}.$$

If we compare this equation with the equation $T = 2\pi \sqrt{\dfrac{l}{g}}$ which gives the time of vibration of a simple pendulum of length l, we see that the compound pendulum has the same time of vibration as a simple pendulum of length l such that $l = \dfrac{h^2 + k^2}{h}$ This is called the length of the simple equivalent pendulum.

As an example of a compound pendulum consider a thin circular ring of radius r resting on a knife edge at its highest point. Its moment of inertia about an axis through its centre and perpendicular to its plane is equal to Mr^2, where M is the mass of the ring. Its centre of gravity is at its centre. Its time of vibration as a pendulum in its own plane is therefore given by

$$T = 2\pi \sqrt{\frac{r^2 + r^2}{gr}} = 2\pi \sqrt{\frac{2r}{g}}.$$

It therefore has the same time of swing as a simple pendulum of length equal to the diameter of the ring. This can be easily verified by hanging a ball from the same knife edge so that it is just level with the bottom of the ring. The two will be found to swing with equal periods.

If we have a body arranged so that it can swing on a number of horizontal parallel axes at different distances from its centre of gravity, then the equation

$$l = \frac{h^2 + k^2}{h}$$

gives the length of the simple equivalent pendulum corresponding to each value of h; and k does not change since the axes are all parallel to the same axis through the centre of gravity. In this case we may regard l as a function of h. The equation giving l may be written

$$\frac{l}{k} = \frac{h}{k} + \frac{k}{h}.$$

The curve in Fig. 52 shows the relation between $\frac{l}{k}$ and $\frac{h}{k}$ given by this equation.

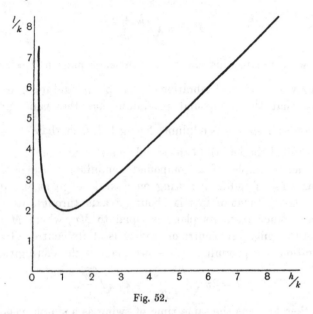

Fig. 52.

We see that when $h/k = 1$ then l/k has its smallest possible value, which is equal to 2. In this case $l = 2k$.

When h/k is very large, k/h becomes very small, so that $l = h$ approximately and the curve nearly coincides with a straight line passing through the origin O.

When h/k is very small we have approximately

$$\frac{l}{k} = \frac{k}{h} \quad \text{or} \quad l = \frac{k^2}{h}.$$

If $h = 0$ then $l = \infty$ so that the time of vibration becomes infinitely long, which means that the body will then stay at rest in any position. For any value of l/k greater than 2 there are two possible values of h/k. Consequently two positions of the axis at different distances from the centre of gravity can be found for which the times of vibration are the same. The equation

$$l = \frac{h^2 + k^2}{h},$$

if solved for h, gives

$$h = \frac{l}{2} \pm \sqrt{\frac{l^2}{4} - k^2}.$$

If h_1 and h_2 denote these two values of h, we have

$$h_1 + h_2 = l$$

and

$$h_1 h_2 = k^2.$$

If then we find experimentally two different values of h which give equal known times of vibration, we can get the length of the simple equivalent pendulum l and the radius of gyration k.

This can be done with a pendulum consisting of a brass bar of rectangular cross section having a series of equidistant holes bored through it. The bar may be about 100 cms. long, 2 cms. wide, and 0·5 cm. thick. The holes should be about 0·6 cm. in diameter and their centres about 2 cms. apart. One end of such a bar is shown in Fig. 53. The bar is supported on a horizontal knife edge put through one of the holes as shown at B. The time of vibration of the bar swinging in its own plane through a small angle can be found by measuring the time of about 100 complete oscillations.

Let the time of vibration be found with the knife edge in each of the holes and also the distances of the points like A, B and C, where the pendulum rests on the knife edge, from one end of the bar. The position of the centre of gravity of the bar is also found by balancing it on the knife edge, and the distance of this from the same end of the bar is measured.

The distances and corresponding times of vibration are then plotted on squared paper and a curve drawn through the points. Such a curve is shown in Fig. 54.

Fig. 53.

The centre of gravity is at G. A line like $ABCD$ cuts the curve in four points for which the times of vibration are equal. The points A and C are at different distances from the centre of gravity and on opposite sides of it; so that the distance from A to C is equal to $h_1 + h_2 = l$. In the same way $BD = h_1 + h_2 = l$. By means of the curve we can find AC and BD, and so get l the length of the simple equivalent pendulum corresponding to the time of vibration represented by points on the line $ABCD$. The value of g can then be calculated by the formula

$$T = 2\pi \sqrt{\frac{l}{g}} \quad \text{or} \quad g = 4\pi^2 l/T^2.$$

The lengths $HD = AH = h_1$ and $HB = HC = h_2$ can also be found, and from them the radius of gyration of the bar about the axis

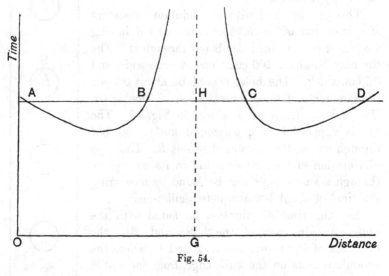

Fig. 54.

through its centre of gravity can be calculated by means of the equation $k^2 = h_1 h_2$. The moment of inertia of a thin uniform rod about an axis through its middle point and perpendicular to it is equal to $M \dfrac{d^2}{12}$, where d is the length of the rod, so that the radius of gyration of the bar will be found to be nearly equal to its length divided by $\sqrt{12}$ or $2\sqrt{3}$.

In the experiment just described the centre of gravity of the bar is always below the knife edge, so that when the knife edge is in the holes on one side of the centre of gravity one end of the bar is at the top and when the knife edge is in the holes on the other side of the centre of gravity the other end is at the top. Care should be taken to measure all the lengths from the same end of the bar. A pendulum which can be inverted like this bar is called a reversible pendulum. The most accurate way of finding the value of g is by means of a form of reversible pendulum invented by Captain Kater and therefore known as Kater's pendulum.

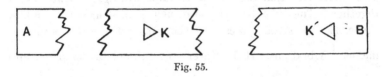

Fig. 55.

For a detailed description of the methods of finding g accurately with various forms of Kater's pendulum, Poynting and Thomson's *Properties of Matter* may be consulted. It will suffice here to describe a simple form of Kater's pendulum capable of giving the value of g correct to within one part in 2000. This consists of a brass bar AB (Fig. 55), about 115 cms. long, having two knife edges K and K' fixed into it, one near one end and the other about 99·3 cms. from the first. These edges are fixed at right angles to the bar and facing each other, as shown in Fig. 55. Between A and K a weight of several pounds is fixed to the bar so as to bring the centre of gravity much nearer to K than to K'. Between K and K' there is a small weight which can be slid along the bar and fixed in any desired position with a screw. The time of vibration of the pendulum should be very nearly 2 secs. on either knife edge. One of the knife edges KK (Fig. 56) is supported on a firmly fixed U-shaped horizontal plane PP and the pendulum set swinging through a small angle. The time of

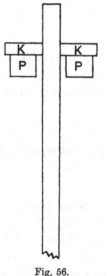

Fig. 56.

vibration is found by comparing it with the time of vibration of the pendulum of an accurate clock. This is best done by a method known as the method of coincidences. The pendulum is put up in front of a clock, having a pendulum with period of 2 secs., so that its lower end is at the same level as that of the clock pendulum and swings parallel to it. The time by the clock is noted when the two pendulums swing exactly together, this is called the time of a coincidence. If they continue to swing together for a long time, say an hour, then their times of vibration are equal. If one gains on the other, then when it has gained a complete vibration the two pendulums will again swing exactly together. The time between two such coincidences is found by the clock, let it be t secs. In the time t the clock pendulum makes $\dfrac{t}{2}$ vibrations. If the Kater's pendulum loses compared with the clock then it makes $\dfrac{t}{2} - 1$ vibrations in the time t. The time of vibration T of the Kater's pendulum is therefore given by

$$T = \frac{t}{\dfrac{t}{2} - 1} = \frac{2t}{t - 2}.$$

If t is large then a small error in t has very little effect on the value found for T, so that this method enables T to be found very accurately. For example, if $t = 1000$ secs. we get

$$T = \frac{2000}{998} = 2 \cdot 00401 \text{ secs.}$$

If we change t to 1010 secs. we get

$$T = \frac{2020}{1008} = 2 \cdot 00397 \text{ secs.}$$

Thus in this case an error of one per cent. in t only produces an error in T of one in 50,000. Having found the time of vibration of the pendulum on one knife edge we then invert it and find its period on the other. By moving the sliding weight these two periods are made as nearly equal as possible. The distance between the knife edges is then $h_1 + h_2 = l$. This distance is measured accurately and then g can be calculated by the formula $g = 4\pi^2 l / T^2$. To save time it is a good plan first to measure the distance between the knife edges and then calculate T, assuming the most probable value of g. Then adjust the sliding weight until the observed value of T for one knife edge agrees with that calculated. It will then be found that the period on the other knife edge has a nearly equal value.

Balance. The beam of a balance is another example of a rigid body which can turn about a horizontal axis. The beam carries three knife edges at right angles to its length, one in the middle turned downwards and one at each end turned upwards. The middle knife edge rests on a horizontal plane and the centre of gravity of the beam is just below it when the beam is horizontal. The pans are hung from planes which rest on the end knife edges. The three knife edges should be in the same plane. Let the distance from the middle knife edge to the centre of gravity of the beam be h and let the radius of gyration of the beam about the middle knife edge as axis be k. Let the mass hung from each of the end knife edges be m and the distance from an end edge to the middle one be d. The masses hung from the end edges increase the moment of inertia by $2md^2$. The time of oscillation of the beam is therefore given by the equation

$$T = 2\pi \sqrt{\frac{k^2}{gh} + \frac{2md^2}{Mgh}},$$

where M is the mass of the beam. If the beam is turned from its equilibrium position through a small angle θ, the restoring couple is equal to $Mgh\theta$. If the mass suspended from one end knife edge is m and that from the other m', then the moment of the weights of these masses about the middle knife edge is

$$mgd - m'gd.$$

The equilibrium position of the beam will therefore be deflected through an angle θ given by

$$Mgh\theta = gd(m - m') \quad \text{or} \quad \theta = \frac{d(m - m')}{Mh}.$$

Thus θ will be large for a small value of $m - m'$ if h and M are small and d large. But if h is made very small the time of vibration of the beam becomes very long; also if M is made very small the beam becomes too weak to support the load without bending appreciably.

In the design of a sensitive balance the beam is designed to be as light and as small as possible, thus keeping M and K small, so that h can be made very small without making the period

Fig. 57.

inconveniently long.   Making d large requires M to be increased so much that no gain in sensitiveness results. A sensitive balance is shown in Fig. 57.

If the middle knife edge is not exactly half-way between the two end ones, then the masses which balance each other are not exactly equal. Let the distances from the middle edge to the end edges be d and d'. Then if masses m and m' balance each other we have $md = m'd'$. If now the mass m is put in the other pan and is then balanced by a mass m'' we have

$$m''d = md'.$$

Hence

$$\frac{m}{m'} = \frac{m''}{m} = \frac{d}{d'},$$

and therefore $m = \sqrt{m'm''}$, and $\dfrac{d}{d'} = \sqrt{\dfrac{m''}{m'}}$.

In this way the ratio $\dfrac{d}{d'}$ can be found and also the mass m if m' and m'' are known masses, e.g. standard weights.

The mass m can also be found by another method. Suppose masses m and m' balance each other. If standard weights are substituted for m so that they also balance m', then these weights are of equal mass to m. In this way m can be found whether d and d' are equal or not.

When a body rotates about a fixed axis it is kept from moving
Balancing Rotating Bodies. in any other way by the axis, so that there are in general forces on the axis and equal and opposite forces on the body. The directions of these forces rotate with the body so that they tend to shake the axis. For example, the rotating parts of machinery may tend to shake the bearings supporting them. In practice the shafts carrying rotating parts cannot be absolutely fixed, so unless the rotating parts are made in such a way that they do not exert appreciable forces on the shafts some shaking results. Such shaking is bad for the building and machinery besides being inconvenient. The rotating parts of machinery are therefore always arranged so that they exert as little force as possible on the shafts carrying them. They are then said to be balanced. Suppose a rigid body is rotating with uniform angular velocity ω about an axis O (Fig. 58) perpendicular to the plane of the paper. Consider a particle P, of mass m in the body, at a distance r from the axis. The force on the axis due to this particle is equal to mv^2/r, where v is the velocity of the particle. But $v = r\omega$, so this force is equal to $mr\omega^2$ and acts

along OP. Take any plane AOB containing the axis and let
the angle between OP and the plane be θ. The perpendicular
distance from P on to this plane is then $r \sin \theta$. Let $r \sin \theta = x$.
Now $\omega^2 mr \sin \theta = \omega^2 mx$ is the resolved part of the force $\omega^2 mr$
perpendicular to the plane AOB. There will be a force like
$mr\omega^2$ for each particle in the body. The sum of all their resolved
parts perpendicular to the plane AOB may be denoted by $\omega^2 \Sigma mx$.
If the centre of gravity of the body lies in the plane AOB we
have $\Sigma mx = 0$, so that then the resultant force on the axis perpendicular to the plane AOB is zero. If the centre of gravity is
on the axis, then in the same way we can show that the resultant
force perpendicular to any plane containing the axis is zero so
that there is then no resultant force on the axis. To balance
a rotating part it must therefore be made so that its centre of

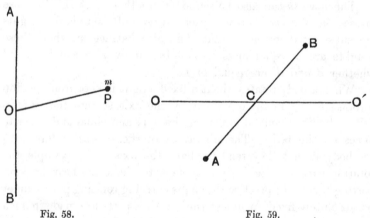

Fig. 58.   Fig. 59.

gravity lies on the axis about which it rotates. There is then
no resultant force on the axis due to its rotation but there may
still be a couple. For a couple has a zero resultant.

For example, let OO' (Fig. 59) be the axis of rotation and let
the rotating body consist of two balls A and B, of equal mass,
attached to a rod AB whose middle point C is fixed to the axis.
The centre of gravity of this body is at C so that there will be no
resultant force on the axis when the body rotates. But the force
due to B and the force due to A will form a couple tending to
turn the axis round from its position OO'. If ACB were at right

angles to OO' there would be no couple.   Thus in order to balance a rotating body it is necessary to balance the couple it exerts on the axis besides making the resultant force zero by having its centre of gravity on the axis.   The couple can always be balanced by adding two equal masses at equal distances from the axis and both lying in the same plane as the axis like the two balls A and B, so as to produce a new couple equal and opposite to that due to the body.

A rigid body rotating round a fixed axis has kinetic energy.

Energy of Rotating Bodies.   Let its angular velocity be ω, and consider a particle in it of mass m and at a distance r from the axis.   The velocity of this particle is $r\omega$; so its kinetic energy is equal to $\frac{1}{2}mr^2\omega^2$.   The total kinetic energy of the body is therefore equal to $\frac{1}{2}\omega^2\Sigma mr^2$ or $\frac{1}{2}K\omega^2$, where K denotes its moment of inertia about the axis.

If a force f, the direction of which lies in a plane perpendicular to the axis, acts on the body, then when the body turns through a very small angle θ the work done by the force on the body is equal to $fp\theta$, where p is the length of the perpendicular between the axis and the direction of the force.   The work done is therefore equal to the moment of the couple acting on the body multiplied by the angle through which the body turns.   If the force f turns round with the body so that fp remains constant, then the work done is equal to $fp\theta$ for any value of θ large or small.

For example, suppose a mass M (Fig. 60) is hung from a wheel, mounted on a horizontal shaft, by a

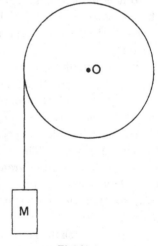

Fig. 60.

thread wrapped round it.   The moment of the tension T in the thread about the axis of rotation is Tr, where r is the radius of the wheel.   If the wheel turns through an angle θ, the work done on the wheel by the tension is $Tr\theta$; and $r\theta$ is equal to the distance through which the weight moves down; so that $Tr\theta$ is equal to the force acting on the wheel multiplied by the distance through which it acts. If K be the moment of inertia of the wheel we have $Tr = K\alpha$, where α is the angular acceleration. The acceleration of the weight is $r\alpha$ so that

$$Mg - T = Mr\alpha$$

or

$$T = Mg - Mr\alpha.$$

Therefore

$$(Mg - Mr\alpha)r = K\alpha,$$

or

$$Mgr = (K + Mr^2)\alpha.$$

Let the wheel start from rest and move through an angle θ and let its angular velocity then be ω.

Then we have

$$\theta\alpha = \tfrac{1}{2}\omega^2,$$

which gives

$$\theta\,\frac{Mgr}{K + Mr^2} = \frac{1}{2}\,\omega^2,$$

or

$$\theta Mgr = \tfrac{1}{2}K\omega^2 + \tfrac{1}{2}Mr^2\omega^2.$$

But $Mgr\theta$ is the work done by the weight so that we see that this is equal to the total kinetic energy of the wheel and the mass—as it should be according to the principle of the conservation of energy.

If a constant couple of moment C acts on a body which can

Power of Couples.

rotate about an axis perpendicular to the plane of the couple, then the work done by the couple when the body turns through an angle θ is $C\theta$. If the body rotates with angular velocity ω, the couple does work equal to $C\omega$ per unit time. The power of the couple is therefore equal to $C\omega$. Power is often transmitted by means of rotating shafts. If a shaft makes n turns in unit time its angular velocity is $2\pi n$, so that the power it transmits is $2\pi nC$ where C is the couple driving it.

Suppose a shaft is driven by a belt passing over a pulley on the shaft (Fig. 61). Let the tension in the belt on one side of the pulley be T and that on the other side T'. Then the moment of the tensions about the axis of the shaft is $r(T - T')$, where r is the radius of the wheel. The power transmitted is therefore $2\pi nr(T - T')$. For example, if $r = 2$ feet, $T = 105$ pounds weight and $T' = 5$ pounds weight, and n is 250 turns per minute, the power is $2\pi \times 250 \times 100 \times 2 = 314{,}159$ foot-pounds per minute. This is equal to

$$\frac{314159}{33000} = 9\text{·}52 \text{ horse power.}$$

Power is often transmitted from one shaft to another by means of two pulleys and a belt. If one pulley has a radius r and makes n revolutions in unit time and the other has a radius r' and makes n' revolutions in unit time, then if the belt does not slip we have $nr = n'r'$, because the same length of belt passes over

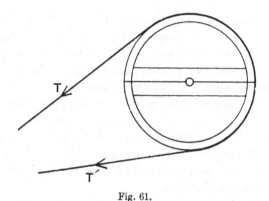

Fig. 61.

each pulley. If C is the couple in the first shaft and C' that in the second, then if there were no loss of power we should have

$$Cn = C'n',$$

and therefore $\qquad\qquad Cr' = C'r.$

The couple required to transmit a given amount of power along a shaft is inversely proportional to the rate of revolution of the shaft.

The equations obtained for rotating bodies may be compared Comparison of with those for particles moving along straight Equations. lines. We have

$$s = tv, \qquad\qquad \theta = \omega t,$$

$$a = \frac{v_2 - v_1}{t}, \qquad\qquad \alpha = \frac{\omega_2 - \omega_1}{t},$$

$$s = \tfrac{1}{2}(v_1 + v_2)\,t, \qquad\qquad \theta = \tfrac{1}{2}(\omega_1 + \omega_2)\,t,$$

$$as = \tfrac{1}{2}(v_2^2 - v_1^2), \qquad \theta\alpha = \tfrac{1}{2}(\omega_2^2 - \omega_1^2),$$

$$f = ma, \qquad\qquad C = K\alpha,$$

$$f = \frac{mv_2 - mv_1}{t}, \qquad C = \frac{K\omega_2 - K\omega_1}{t},$$

Kinetic energy $= \tfrac{1}{2}mv^2$, Kinetic energy $= \tfrac{1}{2}K\omega^2$,

$$W = fs, \qquad\qquad W = C\theta,$$

$$fs = \tfrac{1}{2}m(v_2^2 - v_1^2), \qquad C\theta = \tfrac{1}{2}K(\omega_2^2 - \omega_1^2),$$

$$P = \frac{W}{t} = fv, \qquad\qquad P = C\omega.$$

We see that moment of inertia K takes the place of mass

Angular Momentum. m, and couple C that of force f. Momentum mv is replaced by $K\omega$. This quantity $K\omega$ therefore may be called the *angular momentum* of the body. It is also sometimes called the moment of momentum. Since $C = K\alpha$ we see that if $C = 0$ or if the moment of the forces acting on the body about its axis of rotation is zero, then its angular acceleration is also zero. If $\alpha = 0$ then it follows that ω and therefore $K\omega$ is constant. The angular momentum of a rigid body about any axis therefore remains unchanged so long as no forces act on it having any moment about that axis.

If a heavy wheel is mounted on a shaft which is free to turn in

Gyroscope. any direction about the centre of gravity of the wheel, then if the wheel is made to rotate rapidly it is found that the direction of its axis remains fixed. An apparatus of this kind is shown in Fig. 62. It is called a *gyroscope*. The wheel W is mounted on a shaft AB which is supported by a ring CD. This ring is supported by two bearings at C and D in a frame which can turn about a vertical axis in a bearing at F. The wheel can turn about the vertical axis through F and also about the two perpendicular horizontal axes CD and AB, so that it is free to turn in any direction. If the wheel is set spinning rapidly about its axis AB, then the direction of AB will remain parallel to itself when the apparatus is moved about in any manner.

If now a weight is hung from the ring $CADB$ at A so that it tends to turn the rotating wheel about CD, it is found that the wheel does not turn about CD but moves round about the vertical axis through F, which is perpendicular to CD.

The same thing can be illustrated with a spinning top shown in Fig. 63. This consists of a wooden disk, with an axle C, which is supported at its centre of gravity on a stand S. The top of the stand carries a small sphere which fits into a cup fixed to the disk so that the centre of gravity of the disk is at the centre of the sphere. If the disk is set rotating about its axle, then the direction of the axle remains fixed. If now the rotating disk is pressed down on one side, say at A, by means of a pulley P mounted on a bearing, it is found that the disk resists the pressure and turns about a line AO joining the point of contact of the pulley to the centre of the disk; that is, it turns about an axis perpendicular to the axis about which the pressure applied would turn it if it were at rest.

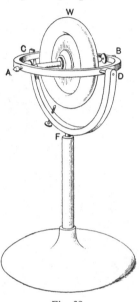

Fig. 62.

Fig. 64 shows the top of the disk. Let it be rotating in the direction shown by the arrow. If it is pressed down at A then the top end of the axle C moves towards F. If it is pressed down at E the top end of the axle C moves towards A. This at first sight surprising property of rotating bodies can be explained simply by means of the equation

$$C = K\alpha \quad \text{or} \quad C = \frac{K\omega_2 - K\omega_1}{t},$$

according to which a couple C acting on the body for a time t increases its angular momentum about an axis perpendicular to the plane of the couple by an amount Ct.

If the body has angular momentum before the couple C acts on it, then the effect of the couple will be to compound with the initial angular momentum an amount Ct about an axis perpendicular to the plane of the couple. For example the top at first has angular

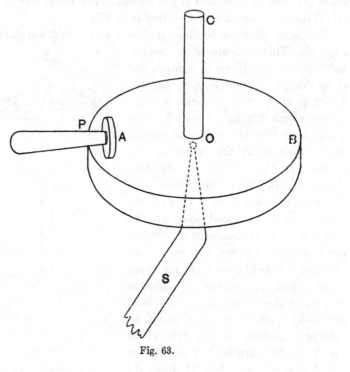

Fig. 63.

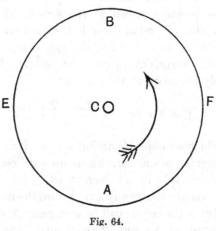

Fig. 64.

momentum about the axis OC. Pressing it down at A gives it angular momentum about the axis EF which is perpendicular to OC. We can show that the effect of this is to turn OC from its original direction towards F.

Angular velocity has magnitude and direction. Its direction is the direction of the axis of rotation. It may therefore be represented by a straight line drawn parallel to the axis and of a length proportional to its magnitude. The line is drawn from the body so that on looking back along it towards the body the body appears to be rotating in the opposite direction to the hands of a watch.

The angular momentum $K\omega$ is proportional to the angular velocity, so it also can be represented in magnitude and direction by a straight line. Angular momentum is therefore a vector and the resultant of the two angular momenta can be found in the same way as the resultant of any two vectors, for example two forces or two displacements. Let OP (Fig. 65) represent the angular momentum of the top before it is pressed down at A. OP is then parallel to the axis of rotation of the top. When it is pressed down at A it is given angular momentum about the perpendicular axis EF. Let this additional angular momentum be represented by OQ. The resultant angular momentum is then represented by the diagonal OR of the parallelogram $OPRQ$. Thus the axis of rotation moves from OP to the direction OR. We have $OQ = Ct$. If the angle POR is small it is equal to PR/OP.

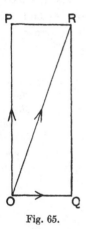

Fig. 65.

Let $K\omega$ be the original angular momentum represented by OP and let $P\hat{O}R = \theta$. Then we have

$$\theta = \frac{Ct}{K\omega}, \text{ or } \frac{\theta}{t} = \frac{C}{K\omega}.$$

If the force exerted at A is f and $AO = r$ we have $fr = C$ so that

$$\frac{\theta}{t} = \frac{fr}{K\omega}.$$

θ/t is the angular velocity with which the top end of the axis of rotation turns towards F about AB as axis when A is pressed down with the force f. This angular velocity may be made very small by making ω very large.

Suppose the top weighs 10 pounds and has a radius of one foot. Let its radius of gyration be, say, 8 inches, so that its moment of inertia is $10\,(\tfrac{8}{12})^2 = 4\cdot44$ in pounds and feet. Suppose it makes 30 revolutions per second, so that $\omega = 2\pi \times 30 = 188$. Let f be equal to one pound weight or 32 poundals and be applied 6 inches from the centre of the top. Then we have

$$\frac{\theta}{t} = \frac{32 \times \tfrac{1}{2}}{4\cdot44 \times 188} = \frac{1}{52} \text{ approximately.}$$

Thus the axis will move towards F at the rate of $\tfrac{57}{52}$ degrees per second, since the unit angle in circular measure is about 57 degrees.

REFERENCES

Matter and Motion, J. Clerk-Maxwell.
Mechanics, Cox.
Spinning Tops, Perry.
Elementary Rigid Dynamics, E. J. Routh.
Properties of Matter, Poynting and Thomson.

CHAPTER VII

GRAVITATION

WE have seen that all bodies on the earth are attracted by
Newton's Law the earth so that they fall towards it with an
of Gravitation. acceleration g which varies from 978 cms./sec.^{2} at
the equator to 983 cms./sec.^{2} at the poles. Newton found that the
observed motions of the planets and their satellites round the sun
could be explained by supposing that there is an attraction between
any two bodies proportional to the product of their masses and
inversely proportional to the square of the distance between them.
The attraction between the earth and bodies at its surface is
a particular case of this universal attraction. If the masses of
two bodies are m_1 and m_2 and their distance apart r, then the
force F with which they attract each other is given by the
equation

$$F = G \frac{m_1 m_2}{r^2},$$

where G is a constant. As an example of this let us compare the
attraction between the moon and the earth with the attraction
between the earth and a body at its surface.

The acceleration a which the force F produces in the body
of mass m_1 is given by

$$G \frac{m_1 m_2}{r^2} = m_1 a,$$

or

$$a = G \frac{m_2}{r^2}.$$

It can be shown that a spherical body of uniform density
attracts a body outside it as though all the matter in it were
concentrated at its centre. If then r_1 denotes the radius of the

earth, r_2 the distance between the earth's centre and the moon's, and m the earth's mass, we have

$$a_1 = G \frac{m}{r_1^2},$$

$$a_2 = G \frac{m}{r_2^2},$$

where a_1 denotes the acceleration with which a body falls to the earth at its surface and a_2 the acceleration of the moon towards the earth. But $a_1 = g$ so that

$$\frac{a_2}{g} = \frac{r_1^2}{r_2^2}.$$

We have $g = 980 \dfrac{\text{cms.}}{\text{sec.}^2}$, $r_1 = 6\cdot37 \times 10^8$ cms., and $r_2 = 3\cdot84 \times 10^{10}$ cms.

Hence $\qquad a_2 = 980 \times \left(\dfrac{6\cdot37 \times 10^8}{3\cdot84 \times 10^{10}}\right)^2 = 0\cdot27 \dfrac{\text{cms.}}{\text{sec.}^2}.$

Now the moon moves round the earth nearly in a circle of radius $r_2 = 3\cdot84 \times 10^{10}$ cms. and its period of revolution $T = 39343$ minutes. Hence its acceleration towards the earth is given by

$$\frac{v^2}{r_2} = \frac{4\pi^2 r_2}{T^2} = \frac{4\pi^2 \times 3\cdot84 \times 10^{10}}{(39343 \times 60)^2} = 0\cdot272 \frac{\text{cms.}}{\text{sec.}^2}.$$

This agrees almost exactly with the value calculated by assuming the acceleration to be inversely proportional to the square of the distance from the centre of the earth. Thus we see that the attraction between the moon and the earth is of the same kind as the attraction between the earth and bodies at its surface.

In the equation $g = G \dfrac{m}{r_1^2}$ the values of g and r_1 are known,

Cavendish's Experiment. so that if we knew G we could get m the mass of the earth. To find G it is necessary to measure the force with which two bodies of known masses attract each other when they are at a known distance apart. This was first done by Cavendish about the year 1797. The following is a description of an apparatus, similar in principle to that used by Cavendish, which enables a rough determination of the constant G and so of the mass of the earth to be done in a few minutes as a lecture experiment.

A wire AB (Fig. 66) about 5 cms. long has at each end a small silver sphere weighing about one gram. This is supported by a wire W which carries a mirror M and the whole is hung up by a fine fibre E, made of fused quartz, which is about 60 cms. long. This apparatus is enclosed in a case. The part of the case surrounding AB is a glass box with double walls and is only wide enough to allow the spheres to move horizontally through a few millimetres. The wire AB can oscillate about a vertical axis

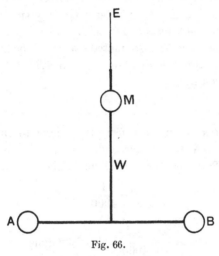

Fig. 66.

under the action of the couple exerted on it by the quartz fibre. If m is the mass of each sphere and $2d$ the distance between the centres of the spheres, then the moment of inertia about the axis of rotation is approximately $2md^2$ so that the time T of a complete oscillation is given by

$$T = 2\pi \sqrt{\frac{2md^2}{C}} \quad \text{or} \quad C = \frac{8\pi^2 md^2}{T^2},$$

where C is the couple exerted by the fibre when it is twisted through unit angle in circular measure. Suppose $d = 2\cdot5$ cms., $m = 1$ gram and $T = 600$ secs., then we get

$$C = \frac{4\pi^2 \times 2 \times 2\cdot5^2}{600^2} = 1\cdot4 \times 10^{-3}.$$

Suppose a force f dynes acts on the sphere at A in a horizontal direction perpendicular to AB and an equal and opposite force acts on the sphere at B. The couple due to these forces is $5f$. If this couple turns AB round through an angle θ, we have

$$5f = C\theta = 1\cdot4 \times 10^{-3}\theta.$$

If $\theta = \dfrac{1}{57\cdot3}$ or one degree, then f is given by

$$f = \frac{1\cdot4 \times 10^{-3}}{5 \times 57\cdot3} = \tfrac{1}{2} \times 10^{-5}\,\text{dyne}.$$

Thus we see that the arrangement described provides an extra-ordinarily sensitive means of measuring very small forces. The angle turned through by AB is found by reflecting a beam of light from the mirror M on to a graduated scale. If the spot of light on the scale moves through a distance s, then AB has turned through an angle θ given by

$$\theta = \frac{s}{2D},$$

where D denotes the distance from the mirror to the scale. The scale is supposed to be perpendicular to the beam of light. Suppose $D = 200$ cms. and $s = 0\cdot1$ cm. Then we have

$$\theta = \frac{1}{4000},$$

so that

$$f = \frac{1\cdot4 \times 10^{-3}}{5 \times 4000} = \tfrac{1}{2} \times 10^{-7}\,\text{dyne}.$$

Thus with this apparatus we can detect a force on the spheres equal to one ten-millionth part of a dyne. The arrangement described is called a torsion balance. To use this apparatus to measure the attraction between two bodies we put up, on either side of the box containing the suspended spheres A and B, two large lead spheres each about 8 cms. in diameter. The positions of these large spheres are shown in Fig. 67. AB is the wire and small spheres in the box CD. The quartz fibre is perpendicular to the plane of the paper at O. The large spheres E and F are carried by parallel horizontal rods GH and MN on which they can slide freely. The sphere E is first put exactly opposite the small sphere A and the sphere F is put opposite B. The attractions between E and A and between F and B cause AB to turn through a small angle and it oscillates about its equilibrium position. The oscillations of the spot of light on the scale are observed, and its position of equilibrium found. The spheres E and F are then moved so that E is opposite B and F opposite A. The attractions between F and A and E and B then tend to turn AB the opposite way, and it moves round and oscillates about a new position. The new equilibrium position of the spot of light on the scale is found. Let s denote the distance between the two equilibrium positions of the spot of light. Moving the large spheres turned

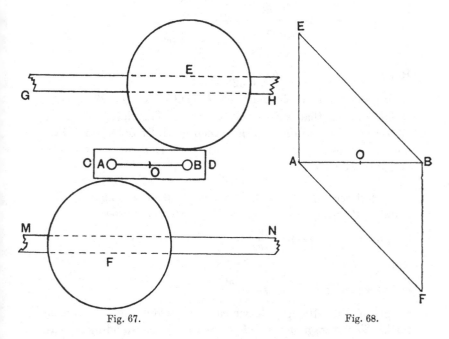

Fig. 67.   Fig. 68.

AB round through an angle $s/2D$ and therefore the couple acting on AB was changed by $Cs/2D$. Let the centres of the large spheres be at E and F and those of the small ones at A and B (Fig. 68). Suppose $EA = AB = BF = 2d$ cms.

Let the mass of each of the spheres E and F be M. The couple on AB due to the attractions between B and F and between E and A is

$$G \frac{Mm}{(2d)^2} 2d = G \frac{Mm}{2d}.$$

There is also a couple in the opposite direction due to the attractions between E and B and A and F. This couple is equal to

$$G\,\frac{Mm}{8d^2} \times 2d\,\frac{\sqrt{2}}{2} = G\,\frac{Mm}{5\cdot 66 d}.$$

The total couple C is therefore given by

$$C = G\,\frac{Mm}{d}\left\{\frac{1}{2} - \frac{1}{5\cdot 66}\right\} = G\,\frac{Mm}{d} \times 0\cdot 323,$$

and twice this is equal to

$$Cs/2D = \frac{s}{2D}\,\frac{8\pi^2 md^2}{T^2},$$

so that

$$\frac{8\pi^2 md^2}{T^2}\,\frac{s}{2D} = 2G\,\frac{Mm}{d} \times 0\cdot 323.$$

Hence

$$G = \frac{2\pi^2}{0\cdot 323}\,\frac{sd^3}{T^2 DM}.$$

If $M = 3000$ grams, $d = 2\cdot 5$ cms., $T = 600$ secs., $D = 200$ cms., it will be found that s is about 15 cms. This makes G about $6\cdot 7 \times 10^{-8}$. The value of G has been carefully determined by a number of observers using a variety of methods. The most probable value of it is

$$G = 6\cdot 66 \times 10^{-8}.$$

A full account of the different methods of finding G will be found in Poynting and Thomson's *Properties of Matter*.

Putting $g = 980\,\dfrac{\text{cms.}}{\text{sec.}^2},\quad r_1 = 6\cdot 37 \times 10^8$ cms.,

and $G = 6\cdot 66 \times 10^{-8}$

in the equation $g = G\,\dfrac{m}{r_1^2},$

we get $m = 6 \times 10^{27}$ grams, which is therefore the mass of the earth. The average density of the earth is got by dividing its mass by its volume and so is equal to

$$m \div \tfrac{4}{3}\,\pi r_1^3 = 5\cdot 5 \text{ grams per c.c.}$$

Since it is found that the weight of a body at any particular place depends only on its mass and is independent of its condition, it follows that the gravitational attraction between any two bodies depends only on their masses and the distance between them.

REFERENCE

Properties of Matter, Poynting and Thomson.

CHAPTER VIII

ELASTICITY

ELASTICITY is that property of matter in virtue of which force
Definition of   is required to change the shape or the volume of a
Elasticity.   piece of matter and to maintain the change. For
example, if a weight is hung from one end of a piece of india-
rubber cord, the cord becomes longer and thinner. This change
of shape persists so long as the weight is not removed, but when
the weight is removed the indiarubber goes back to its original
shape. If a straight wooden bar is supported horizontally at its
ends and a weight hung from the middle, it becomes bent. When
the weight is removed it springs back to its original shape. Solid
bodies require force to change their shape, but fluid bodies do not
resist a change of shape with a permanent force. While the
shape of fluid bodies is being changed there are forces which
resist the change, but these disappear when the change is complete and are smaller the smaller the rate at which the change is
made. If the change of shape is made extremely slowly, then
with fluids the forces required to produce it become extremely
small, whereas with solids the force required is nearly as great for
a slow change of shape as for a rapid change. Fluids resist a
change of volume with a permanent force and spring back to their
original volume when the force is removed. Solid bodies may be
said to possess elasticity of shape and of volume, while fluids
possess no elasticity of shape but only elasticity of volume.

When the shape or size of a body is altered it is said to
                 be strained and the change of shape or size is called
Strain.   a strain.

When a body is first strained by the application of forces to it
and then the forces are removed, if the body then springs back to

its original size and shape so that the strain completely disappears, the body is said to be perfectly elastic for the strain in question. Fluids are found to be perfectly elastic for changes of volume.

If the volume of a body is V and it is changed to V' without any change of shape, then the strain is taken equal to $\dfrac{V' - V}{V}$, that is to the change of volume per unit volume. For example, if a sphere of radius r is changed unto a sphere of radius r', by compressing it equally in all directions the strain is

$$\frac{\frac{4}{3}\pi r'^3 - \frac{4}{3}\pi r^3}{\frac{4}{3}\pi r^3} = \frac{r'^3 - r^3}{r^3}.$$

If V' is less than V the strain is negative. A change of volume without change of shape is called a uniform dilatation.

If a rectangular block $ABCDEFGH$ (Fig. 69) is strained by moving the top parallel to the base from $ABCD$ to $A'B'C'D'$ while the base $EFGH$ remains fixed, then it is changed into

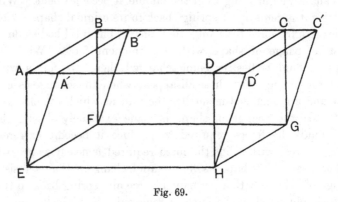

Fig. 69.

a parallelepiped $A'B'C'D'EFGH$, the volume of which is equal to the original volume of the block. This sort of strain alters the shape without altering the volume and it is called a *shear*. A shear is measured by the angle AEA' which is called the angle of shear.

It can be shown that any strain is made up of a dilatation and a shear.

When force is applied to a surface, so that at every point on the surface the direction of the force there is

Pressure. perpendicular to the surface, then the surface is said to be acted on by a pressure. The pressure is taken equal to the force per unit area acting on the surface. Thus if a uniform pressure p acts all over a plane surface of area α the force f on the area is given by $f = p\alpha$.

In order to change the volume of a body without changing its

Bulk Modulus shape, a uniform pressure must be applied all over

of Elasticity. its surface. If a pressure p applied in this way changes the volume of a body from V to V', then it is found that

$$p = -k\frac{V' - V}{V},$$

where k is a constant. A positive pressure diminishes the volume—which is the reason for the minus sign. The constant k is called the bulk modulus of elasticity.

When a plane surface is acted on by a force, uniformly

Tangential distributed over the surface, and everywhere parallel

Stress. to a line on the surface, it is said to be acted on by a uniform tangential stress. If the force per unit area is T, the total force is equal to $T \times$ (area of surface).

If the four surfaces $ABCD$, $EFGH$, $ABFE$, and $DCGH$ of the

Shearing rectangular block $ABCDEFGH$ (Fig. 69) are acted

Stress. on by four equal uniform tangential stresses (T) in the directions A to D, H to E, A to E and H to D respectively, the block will be strained into a parallelepiped like $A'B'C'D'EFGH$, its volume remaining unchanged. It is found, provided the strain is very small, that $T = n\theta$ where θ denotes the circular measure of the angle AEA' and n is a constant which is called the modulus of rigidity.

The following table gives some values of n and k:

Substance	n	k
Steel......	$(7 \text{ to } 10) \times 10^{11}$	$(14 \text{ to } 19) \times 10^{11}$
Brass ...	$(3 \text{ to } 4) \times 10^{11}$	$(10 \text{ to } 11) \times 10^{11}$
Glass......	$(1 \text{ to } 3) \times 10^{11}$	$(3 \text{ to } 4) \times 10^{11}$
Water ...	0	$0 \cdot 2 \times 10^{11}$

These are the values of n and k when the stresses are expressed in dynes per sq. cm.

As an example, suppose a cubical block of steel with sides 100 cms. long is acted on by a uniform pressure all over its surface equal to 1000 kilograms weight per sq. cm. This is a pressure of $10^6 \times 980$ or 9.8×10^8 dynes per sq. cm. The change of volume of the block is given by

$$9.8 \times 10^8 = -16 \times 10^{11} \frac{V' - V}{10^6},$$

which gives $V - V' = 610$ cubic centimetres.

A pressure of 1000 kilograms weight per sq. cm. is equal to 14,200 pounds weight per sq. inch. We see then that only very small changes can be produced in the volumes of most solids and liquids even by enormous pressures.

Fluids, as we have seen, are distinguished from solids by the fact that they have no elasticity of shape. Fluids are of two kinds, liquids and gases. Liquids have a definite volume; so that when some liquid is contained in a vessel of volume greater than that of the liquid, the liquid occupies only a portion of the volume of the vessel. A gas always completely fills up the vessel in which it is contained; so that its volume is equal to that of the vessel containing it. The gas exerts a pressure on the walls of the vessel containing it which depends on the volume of the vessel and the amount of gas in it. If the volume is changed from V to V' and the pressure changes from p to p', then if $V' - V$ is very small it is found that

$$p' - p = -k \frac{V' - V}{V},$$

where k is a constant. k is the bulk modulus of elasticity of the gas. The properties of a gas depend greatly on whether it is hot or cold, that is on its temperature, so that it is best to deal with them in the chapters on Heat.

REFERENCES

Properties of Matter, Poynting and Thomson.
Experimental Elasticity, G. F. C. Searle.

CHAPTER IX

THE PROPERTIES OF LIQUIDS

LIQUIDS have no elasticity of shape, but occupy a definite
Pressure in volume. Consider a small area α on the surface
Liquids. of a solid body immersed in a liquid which is at
rest relative to α. The force exerted by the liquid on the area α
must be equal and opposite to the force exerted by the area α on
the liquid. This force must be perpendicular to the area α, for if
not it would have a component parallel to the surface which
would tend to produce a shearing strain in the liquid, and since a
liquid has no elasticity of shape it cannot resist a force tending to
shear it without being set in motion. The force exerted by a liquid
when at rest on any surface in contact with it is therefore normal
to the surface; the pressure on the surface is measured by the
force per unit area. If we consider any small area in the liquid,
then the liquid on one side of it exerts a force on the liquid on the
other side. This force also when the liquid is at rest must be
normal to the area. The force per unit area is called the pressure
in the liquid at the place where the small area is situated. We
can show that this pressure is the same in all directions at any
point in the liquid. Consider a very small triangle ABC (Fig. 70)
in a liquid at rest. Draw AA', BB' and CC' perpendicular to the
plane of the triangle and of equal lengths. Join $A'B'$, $B'C'$ and
$C'A'$. The liquid inside the prism $AA'C'B'BC$ is in equilibrium
under the action of the forces exerted on it by the surrounding
liquid and its weight. If we take the prism small enough, its
weight can be neglected, for its weight is proportional to its
volume while the forces on it, due to the surrounding liquid, are
proportional to the areas of its surfaces. The forces on the
surfaces of the prism must therefore be in equilibrium and they
are normal to the surfaces on which they act and proportional to

the areas. The force on ABC and the opposite force on $A'B'C'$ must balance each other because the other three forces on $ABB'A'$, $AA'C'C$ and $BB'C'C$ are perpendicular to the forces on ABC and $A'B'C'$. The three forces on $ABB'A'$, $AA'C'C$ and $BB'C'C$ must therefore be in equilibrium. But they are perpendicular to the three sides of the triangle ABC, so that if this triangle were turned through a right angle in its own plane, its sides would be parallel to the three forces. But when three forces in

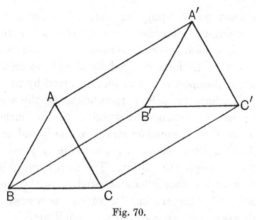

Fig. 70.

equilibrium are parallel to the sides of a triangle, then the sides of the triangle are proportional to the forces. But the sides of ABC are proportional to the areas of $ABB'A'$, $AA'C'C$, and $BB'C'C$, so that the forces on these areas must be proportional to the areas. The force per unit area, or the pressure, is therefore the same on each of the three areas $ABB'A'$, $AA'C'C$, and $BB'C'C$. Since the small prism can be supposed placed in any position, it follows that the pressure at a point in a liquid is the same in all directions.

This fact can be illustrated in the following way. The top of a small cylindrical metal box AB (Fig. 71) is closed by a thin rubber sheet. To the middle of this sheet a wire K is fastened which turns a balanced lever EFG about a bearing at F. The end of the lever moves over a graduated scale CD which is supported by an arm attached to the box. If this apparatus is immersed in water, the pressure of the water on the rubber

membrane pushes it in and so the lever moves over the scale.
It is found that the deflection of the lever is the same at a given
depth in whatever direction the surface of the membrane faces,
which shows that the pressure is the same in all directions. The
deflection obtained depends on the depth to which the box
is immersed. It is greater the greater the depth.

If a pressure is applied to any portion of the surface of a

Pascal's Principle. liquid and the liquid remains at rest, then the
pressure is transmitted throughout the liquid so
that the pressure in it is everywhere increased by an amount
equal to the pressure applied. To prove this, consider a long thin

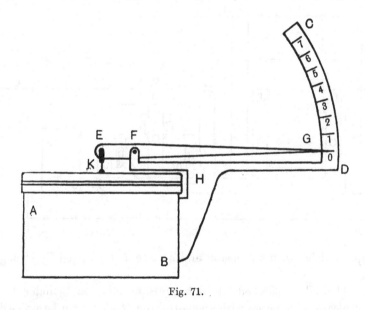

Fig. 71.

straight circular cylinder in a liquid at rest. The force on the
sides of the cylinder is everywhere perpendicular to its axis
and so has no component along the axis. If then the pressure
on one end is increased, the pressure on the other end must
be increased at the same time by an equal amount, or else
the cylinder will be set in motion parallel to its length. Such
cylinders can be imagined drawn in any direction in a liquid
so that in a liquid at rest a pressure is transmitted equally

to all parts of the liquid. This is known as Pascal's principle after its discoverer. This property of liquids is made use of in a machine called the hydraulic press, by means of which very large forces can be obtained.

Fig. 72 shows a hydraulic press. P is a cylindrical piston which can slide up and down in a cylinder CC. Oil or some other liquid is forced into the cylinder through a pipe S. If the pressure in the oil is p and the area of cross section of the piston P is a, then the upward force on the piston is pa. This force drives the piston up, and any body A placed on the top of the piston is

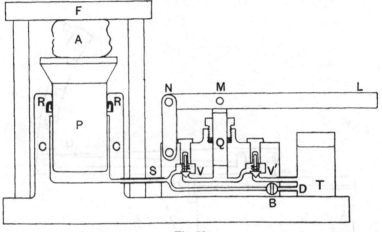

Fig. 72.

squeezed between the piston and a plate F supported by strong bars.

The oil is pumped at high pressure into the cylinder CC by means of a pump with a small piston Q which can be pressed down by a lever LMN. V and V' are two valves consisting of metal cones which fit on to conical surfaces and are held down by springs as shown. When Q is moved down, the valve V rises and allows the oil to pass into the cylinder CC, but it does not allow any oil to flow back. When Q is raised the valve V' rises and allows oil to flow in from a tank T, but it does not allow any to flow back into T when Q is pressed down. Thus when the piston Q is worked up and down, the oil is forced from the tank into the cylinder CC and the piston P rises. If a cock B is opened the oil flows back into the tank through a pipe BD and the piston P moves down. The oil is prevented from escaping round the sides of P by a leather ring RR of U-shaped cross section. The pressure of the oil forces this leather ring against the sides of P so that the oil cannot escape. Let the downward force on the lever at L be denoted by F. Let $NL = d$ and $NM = d'$. The force on the top of the piston Q is therefore Fd/d'. Let the area of cross section of Q be α', then the pressure p in the oil due to the piston Q is given by

$$p = \frac{Fd}{\alpha'd'}.$$

The upward force exerted by the piston P is therefore equal to $Fd\alpha/d'\alpha'$. α can easily be made much larger than α' and d than d', so that this force may be enormously greater than F. For example, suppose P is two feet in diameter and Q one inch in diameter. Then

$$\frac{\alpha}{\alpha'} = (24)^2 = 576.$$

Also suppose d is ten times d'. Then the force exerted by the large piston is 5760 times the force F applied at L. In this case a force equal to 100 pounds weight at L gives a force of 576,000 pounds weight on the body A, or about 260 tons weight. Various types of hydraulic presses are much used in modern engineering practice, and presses capable of exerting a force equal to 10,000 tons weight are in use. Large masses of steel are pressed while red hot into the desired shape by means of the hydraulic press. This method is found much superior to the old plan of using a steam hammer.

The free surface of a liquid at rest in a vessel lies in a

Pressure at different depths. horizontal plane. To prove this consider a particle of the liquid at the surface. The weight of the particle is directed downwards, so that if the surface were not horizontal the weight would have a component parallel to the surface which would set the liquid in motion, for a liquid cannot resist a tangential force. The surface therefore must be horizontal when the liquid is at rest. The air or atmosphere exerts a pressure on all bodies in contact with it which is equal to about 15 pounds weight on the square inch, or one million dynes per sq. cm. The pressure at the free surface of a liquid when exposed to the air is equal to the atmospheric pressure. Let us now consider the pressure in a liquid with a free surface exposed to the air, at different depths below this surface. Let b denote the pressure of the air. In Fig. 73, let EF be a vessel containing any liquid. Imagine a cylinder $ABCD$ with its sides vertical and one end AB at the free surface of the liquid and the other at a depth h below the free surface. The liquid in this cylinder is supported by the forces exerted on it by the surrounding liquid. The forces

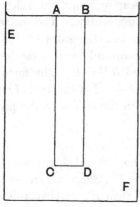

Fig. 73.

on the sides of the cylinder are horizontal and so do not help to support the weight of the cylinder. Let the horizontal cross section of the cylinder be α. Then there is a downward force on the top of it due to the pressure of the air equal to $b\alpha$, and there is an upward force on the bottom of it equal to $p\alpha$, where p is the pressure in the liquid at the depth h.

If w is the weight of the cylinder, we have therefore

$$b\alpha + w = p\alpha$$

or

$$w = (p - b)\,\alpha.$$

Let the mass of unit volume of the liquid be ρ, so that the weight of unit volume in dynamical units of force is ρg. Then since the volume of the cylinder is $h\alpha$, we have $w = h\alpha\rho g$;

hence

$$h\alpha\rho g = (p - b)\,\alpha,$$

or

$$p - b = h\rho g.$$

The mass of unit volume of any substance is called its density. It appears therefore that the difference between the pressure at a point in a liquid at rest and the pressure at its free surface is proportional to the depth of the point and to the density of the liquid. The pressure is therefore the same at all points in any horizontal plane in a liquid at rest. The density of water is nearly equal to one gram per cubic cm., so that the pressure in water at a depth h cms. below its free surface is nearly equal to $980h + b$ dynes per sq. cm. or, since $b = 10^6$ dynes per sq. cm. nearly, it is nearly equal to $980h + 10^6$. For example, at a depth of 100 metres the pressure in the water is $9,800,000 + 10^6 = 10\cdot8 \times 10^6$ dynes per sq. cm. This is about $10\cdot8 \times 15 = 162$ pounds weight per sq. inch.

If two vessels of any shape are connected by a tube and a liquid poured into them, the free surfaces of the liquid in both vessels will be at the same level when the liquid comes to rest provided the tube becomes filled with the liquid. For the pressure in a liquid is constant over any horizontal plane; so that the free surfaces, at which the pressure is equal to atmospheric pressure, must be all in the same horizontal plane. This may be illustrated with the apparatus shown in Fig. 74. The free surfaces at A, B and C are all at the same level whatever the shape of the vessels.

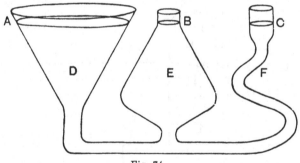

Fig. 74.

The fact that the pressure in a liquid at a given depth is independent of the shape of the vessel containing it can be shown experimentally with the apparatus shown in Fig. 75. Three glass vessels A, B and C, with slightly conical lower ends which fit into a tube T, are provided. A piston P can slide freely up and down in the tube T, and is supported by a lever L carrying a weight W. If the water is poured into A, then when the pressure on the piston reaches a certain value the lever tips up. The depth of the water above the top of the piston when this happens is indicated on a scale S carried by the piston. If now the same

thing is tried with the vessels B and C, it is found that the depths
at which the lever tips are the same with all three vessels, although
the weight of water in A is much greater than that in C and the
weight of water in B is much less than that in C. The forces
exerted on the water in A by the sloping sides of the vessel have
upward components which help to support the weight of the water;
so that the pressure on the piston is the same as when the vessel

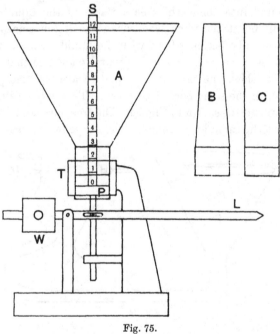

Fig. 75.

C is used. In the same way the forces exerted by the sides of B
have downward components which make the force on the piston
greater than the weight of the water in the vessel.

When a solid body is immersed in a liquid the forces, exerted

Force on body
immersed.

on it by the liquid, have a resultant which is
directed upwards, so that they tend to move it
upwards. If this upward force is greater than the weight of the
body, the body moves up and floats, but if the weight is greater
then the body sinks.

Let PQ (Fig. 76) be a solid body of any shape immersed in a liquid with its free surface at RS. Take a small area AA' equal to α on the free surface, and describe a cylinder $ABCC'B'A'$ with vertical sides and cross section α. Let this cylinder cut the surface of the body at BB' and CC'. Let the area of BB' be α' and that of CC' be α''. Let the angle between a normal to the area BB' and the vertical length of the cylinder be θ' and that between the normal to CC' and the vertical be θ''. Let $AB = h'$ and $AC = h''$. The pressure at B is $h'\rho g + b$, where ρ is the density of the liquid and b the atmospheric pressure. The force on the area BB' is therefore $\alpha'(h'\rho g + b)$ and the component of this vertically downwards is $\alpha'(h'\rho g + b)\cos\theta'$. In the same way the component of the force

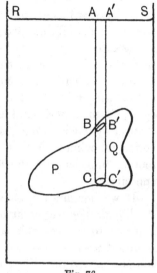

Fig. 76.

on CC' vertically upwards is $\alpha''(h''\rho g + b)\cos\theta''$. But $\alpha'\cos\theta' = \alpha$ and $\alpha''\cos\theta'' = \alpha$ so that the upward force on the body due to the forces on the two areas BB' and CC' is

$$\alpha(h''\rho g + b) - \alpha(h'\rho g + b) = \alpha\rho g(h'' - h').$$

The volume of the cylinder $ACC'A'$ inside the body between B and C is equal to $\alpha(h'' - h')$ so that $\alpha\rho g(h'' - h')$ is equal to the weight, in dynamical units of force, of a volume of the liquid equal to the volume of the cylinder inside the body. The whole of the free surface of the liquid above the body may be supposed divided into small areas like α, and corresponding to each small area there is an upward force on the body equal to the weight of a volume of the liquid equal to the volume of the body vertically below the small area. Thus we see that the total upward force on the body is equal to the weight of a volume of the liquid equal to the whole volume of the body. If the volume of the body is V, then the upward force on it due to the liquid is equal to $V\rho g$.

If the mass of the body is m, then the resultant force on it is equal to $V\rho g - mg$ upwards. If the density of the body is ρ' then $m = V\rho'$, so that the resultant upward force F on it is given by $F = V\rho g - V\rho' g = Vg\,(\rho - \rho')$. Thus if $\rho = \rho'$ the body will be in equilibrium and will not move either up or down. If ρ' is greater than ρ it will sink to the bottom, and if ρ is greater than ρ' it will rise to the surface and float.

So far we have only considered the vertical force on the body. It can easily be shown that there is no resultant horizontal force on it, for if a horizontal cylinder is drawn through it like the vertical cylinder $ACC'A'$, since the pressure is the same in the horizontal cylinder on both sides of the body, we see that the horizontal components of the forces on the two sides are equal and opposite.

If we imagine the body PQ replaced by an equal volume of the liquid, then we see at once that the forces on it are just those required to support this equal volume of the liquid, so that the resultant force on the body due to the liquid is evidently equal and opposite to the weight of an equal volume of the liquid. The line of action of the resultant force on the body therefore passes through the centre of mass of a body of uniform density occupying the same space as the body. If the body is not of uniform density its centre of mass may not be on the line of action of the resultant force on it due to the liquid. In this case it will tend to turn round until its centre of mass does lie on the line of action of the resultant.

The upward force on a body immersed in a liquid may be measured by suspending it from one of the pans of a balance by a fine wire and allowing it to hang in a vessel of the liquid. If its weight is w and the weight required to balance it when completely immersed is w', then $w - w'$ is the upward force on it due to the liquid. We have therefore

$$w - w' = V\rho g,$$

where ρ is the density of the liquid and V the volume of the body. Also $w = V\rho' g$, where ρ' is the density of the body. Hence

$$\frac{w - w'}{w} = \frac{\rho}{\rho'}.$$

If the liquid is water, for which ρ is nearly equal to one gram per cubic cm., we have approximately $\rho' = \dfrac{w}{w - w'}$. The average density of a body in grams per c.c. can be found roughly by dividing its weight by the difference between its weight and its apparent weight in water. But if an accurate value of the density is required, then $\dfrac{w}{w - w'}$ must be multiplied by the density of the water at the temperature it had when the body was immersed in it.

If the density in pounds per cubic foot is required, then in the equation $\rho' = \dfrac{\rho w}{w - w'}$ the density ρ of the liquid must be expressed in pounds per cubic foot. The density of water is about 62·4 pounds per cubic foot.

Floating Bodies. If the weight of a body is less than the weight of an equal volume of a liquid, then the body will float in the liquid. The volume of the body beneath the surface of the liquid when it is floating is such that the weight of an equal bulk of the liquid is equal to the weight of the body. If a body of volume V and average density ρ' floats on a liquid of density ρ, then the volume v immersed is given by the equation $v\rho = V\rho'$. For example the density of sea water in grams per c.c. at 0° C. is 1·026 and the density of ice is 0·917. The fraction of the volume of an iceberg which is below the surface of the sea is therefore $\dfrac{0·917}{1·026}$ or 0·894. The fraction above the surface is therefore $1 - 0·894 = 0·106$. When a ship is loaded it sinks into the water so that the increase in the volume beneath the surface of the water is such that the weight of an equal volume of water is equal to the weight of the load put into the ship.

Force on an immersed surface. Suppose we have a plane rectangular area AB (Fig. 77) immersed in a liquid so that one of its sides at B lies in a horizontal plane and the opposite side at A below this plane. Let $AB = l$ and let the width of the area perpendicular to the paper be b, so that the area is bl. Suppose it is required to find the resultant force on this area due to the liquid. Let the depth at B be $h = BB'$ and the depth at A be

$h' = AA'$. At a point P at a distance $PB = x$ from B the depth is therefore $h + (h' - h)\dfrac{x}{l} = d$. The force on unit area at P is therefore $d\rho g$, where ρ is the density of the liquid.

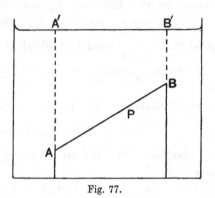

Fig. 77.

Divide the area AB into a large number n of narrow strips of equal width by parallel straight lines drawn across it perpendicular to the plane of the paper. Let these strips be numbered 1, 2, 3, 4, etc., up to n, starting at B. The area of each strip is $\dfrac{lb}{n}$. The force on the 10th strip is

$$\frac{lb}{n}\left(h + (h' - h)\,\frac{10\dfrac{l}{n}}{l} \right)\rho g.$$

The total force on all the strips is therefore

$$lb\rho g\left\{\frac{h}{n}(1 + 1 + \ldots \text{ to } n \text{ terms}) + \frac{h' - h}{n^2}(1 + 2 + 3 + \ldots + n)\right\}$$

$$= lb\rho g\left\{h + \frac{h' - h}{2}\right\},$$

since n is very large,

$$= lb\rho g\,\frac{h + h'}{2}.$$

Thus the force is the same as if the depth all over the area was equal to the average depth. The force of course is perpendicular to the area. To find the point at which the resultant acts we observe in the first place that it must lie on a line parallel to AB down the middle of the area. Also since all the forces are parallel the moment of the resultant about any line must be the same as the total moment of all the forces.

Divide the area into strips as before. The moment of the force on the 10th strip about the top side of the area is

$$\left\{ h + (h' - h) \frac{10\frac{l}{n}}{l} \right\} \rho g \frac{bl}{n} \times 10 \frac{l}{n}.$$

The total moment about this line is therefore

$$l^2 b \rho g \left\{ \frac{h}{n^2} (1 + 2 + 3 + \dots + n) + \frac{h' - h}{n^3} (1^2 + 2^2 + 3^2 + \dots + n^2) \right\}$$

$$= l^2 b \rho g \left\{ \frac{h}{2} + \frac{h' - h}{3} \right\} = l b \rho g \frac{h + h'}{2} . \bar{x},$$

where $\bar{x}$ denotes the distance of the point of action of the resultant from the top side of the area. Hence we get

$$\bar{x} \left(\frac{h + h'}{2} \right) = l \left(\frac{h}{2} + \frac{h' - h}{3} \right).$$

This reduces to $\bar{x} = \dfrac{l}{3} \left(\dfrac{h + 2h'}{h + h'} \right)$. If the top of the area is at the surface of the liquid, then $h = 0$ and $\bar{x} = \dfrac{2}{3} l$. Also if the area is horizontal so that $h = h'$ we get $\bar{x} = \dfrac{l}{2}$, so that then the resultant acts at the middle point of the area.

Since the pressure due to a column of liquid of height h is equal to $h \rho g$, where ρ is the density of the liquid, a column of liquid in a glass tube is often used as a means of measuring pressures. Such an arrangement is called a manometer. A simple form of manometer is shown in Fig. 78. This consists of a glass tube bent as shown and about half full of liquid. Mercury is often used in manometers because it gives off very little vapour, it does not wet the glass, and the position of its surface can be easily seen. A graduated scale is fixed alongside the manometer tube, which should be vertical. If the pressure of the gas in A is greater than the atmospheric

Manometer.

pressure in the open end B, then the liquid stands higher in B than in A. If the difference of level is h, then the difference between the pressure in A and the pressure in B is equal to $h\rho g$. If the end of the tube at B is closed and there is no gas in the tube BC above the liquid, then the pressure there is zero, so that the pressure in A is then equal to $h\rho g$.

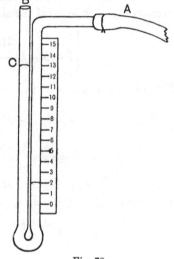

Fig. 78.

In practical work a pressure is often recorded by stating the height of a column of liquid which produces an equal pressure. For example, we may record a pressure as equal to 760 mms. of mercury. Pressures stated in this way can be reduced to dynes per sq. cm. by multiplying by ρg, ρ being the density of the liquid in grams per c.c. and g the acceleration of gravity at the place where the pressure was measured. This method of recording pressures is often convenient, but it has the disadvantage that the pressure due to a column of liquid of given height depends on the temperature of the liquid and on the value of g at the place. Unless the temperature and the value of g are given as well as the height of the column of liquid, it is impossible to find out the exact value of the pressure.

The pressure of the atmosphere is usually measured by means of a special type of mercury manometer called a barometer. A simple form of barometer is shown in Fig. 79. It consists of a glass tube $ACBED$, having wide parts AC and ED, about 2 cms. in diameter, joined by a narrow tube CBE. The parts AC and DE are in the same straight line. The tube is closed at A and open at D. The tube is filled with pure dry mercury, which is boiled in it to expel all air. When it is placed in a vertical position the mercury in AC falls, leaving a vacuum above it. The difference of level between the surface of the mercury in AC and the surface in DE is read off on a

Barometer.

scale HK, which is usually graduated into millimetres. This difference is the height of the barometer. The pressure of the atmosphere is equal to $h\rho g$, where h is the height of the barometer and ρ the density of the mercury. At sea level the average height of the barometer is about 76 cms. when the mercury is at $0°$C. The density of mercury at $0°$C. is 13·596 grams per c.c. If we take $g = 980$ we get the pressure due to 76 cms. of mercury equal to $1·013 \times 10^6$ dynes per sq. cm. The pressure due to 75 cms. of mercury is almost exactly 10^6 dynes per sq. cm.

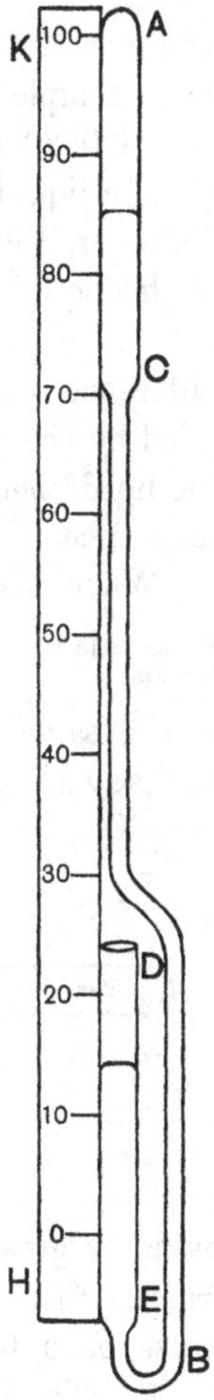

So far we have considered the properties of liquids when at rest or in equi-

Viscosity.

librium. We shall now consider the motion of liquids in a few simple cases. Consider two large parallel plane surfaces at a distance d apart, and let the space between them be filled with any liquid. Suppose one of the planes is held fixed and the other moved parallel to itself in a straight line with uniform velocity v. The distance between the plane surfaces then remains constant. It is found that the presence of the liquid between the two planes causes a force on the moving plane proportional to its velocity and in a direction parallel to its surface and opposite to the direction in which it is moving. This force is spread uniformly over the area of the plane so that it is a uniform tangential stress. Let the force be equal to T per unit area, then it is found that $T = -\mu\dfrac{v}{d}$, where μ is a constant which is called the viscosity of the liquid. If the two planes are unit distance apart and one of them moves with unit velocity parallel to itself, then the force on it per unit area is equal to the viscosity of the liquid.

Fig. 79.

The following table gives some values of μ for liquids at a temperature of $0°$ C. when the force is measured in dynes and the centimetre and second are the units of length and time.

Substance	Viscosity
Water	0·018
Ethyl Alcohol	0·018
Ether	0·0029
Mercury	0·017

The viscosity of most liquids gets less as their temperature rises. For example the viscosity of water at 15° C. is 0·0115, at 30° C. it is 0·0080 and at 45° C. it is 0·0060.

If a liquid is flowing at a constant rate along a pipe of uniform diameter, there is a resistance to the flow due to the viscosity of the liquid. In consequence of this the pressure in the liquid falls as we pass along the pipe in the direction of the flow. The difference between the pressures at two points in the pipe multiplied by the area of cross section of the pipe is the force driving the liquid between the two points and is balanced by the viscous resistance.

Pressure in a Stream. When the velocity with which a stream of liquid moves is changing, so that the liquid has an acceleration, then force is required to produce the acceleration in the same way as for any other body. For example, suppose a steady stream of liquid is flowing through a pipe AB (Fig. 80)

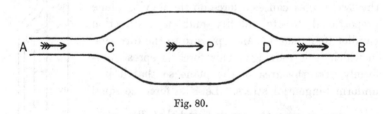

Fig. 80.

having a greater diameter between C and D, as shown. In going from A to P the velocity falls since the tube gets wider, so there must be a force acting on the liquid in the direction from P to A. The pressure in the liquid at P must therefore be greater than the pressure at A. In the same way the velocity at B is greater than at P; so that the pressure at P must be greater than at B. Thus it appears that in a stream of liquid the pressure is greater at places where the velocity is smaller.

If a liquid issues from an aperture in the side of a vessel
Flow from   containing it, then if the pressure inside the vessel
Apertures.   near the aperture is greater than the atmospheric
pressure by an amount p, the work done on the liquid by this
excess of pressure is pV, where V is the volume of liquid which
comes out. If the velocity of the issuing jet of liquid is v, then
its kinetic energy is $\frac{1}{2}mv^2$, where m is its mass. The work done
on it is nearly equal to its kinetic energy, so that we have

$$pV = \tfrac{1}{2}mv^2.$$

Also $m = V\rho$, where ρ is the density of the liquid, so that

$$p = \tfrac{1}{2}\rho v^2.$$

If the pressure p is due to a depth of liquid h, then $p = h\rho g$,
so that

$$h\rho g = \tfrac{1}{2}\rho v^2,$$

or $$v = \sqrt{2gh}.$$

If a body falls freely from a height h its velocity at the bottom
is equal to $\sqrt{2gh}$; so that it appears that when a liquid issues from
an aperture in the side of a vessel at a depth h below the free
surface in the vessel the velocity of the issuing jet of liquid is
nearly equal to the velocity acquired by a body in falling freely
a height h. Some of the work done on the liquid as it comes out
is used up in overcoming the viscous resistance to the motion
of the liquid so that the velocity is always less than $\sqrt{2gh}$. If
a very viscous liquid is used the velocity may be very small, but
with water issuing from a fairly large hole it is nearly equal to
$\sqrt{2gh}$. Let BC (Fig. 81) be a vessel of water with a hole at A.
Let $FA = h$. Then the water comes out at A with a horizontal
velocity $\sqrt{2gh}$. Let $AD = d$ and $DE = l$. Let t be the time a
particle of the water takes to fall from A to E. Then since the
initial velocity at A is horizontal, the time t is equal to the time
a body would take to fall vertically from A to D starting from
rest. Hence $t = \sqrt{\dfrac{2d}{g}}$. Also $l = \sqrt{2gh} \cdot t$, since while the water
is falling it moves horizontally with velocity $\sqrt{2gh}$. Hence

$$l = (\sqrt{2gh}) \times \left(\sqrt{\dfrac{2d}{g}} \right) = 2\sqrt{dh}.$$

If A is half-way between F and D so that $d = h$, we get $l = 2h$. It is easy to measure the distances h, d and l, so that the formula $l = 2\sqrt{dh}$ can be verified experimentally.

If the area of the aperture from which the liquid issues is α, then it is found that the volume coming out in unit time is less than $v\alpha$, where $v = \sqrt{2gh}$ is the velocity of the jet. This is due to the cross section of the jet being less than that of the aperture. Close to the aperture the liquid moves towards the axis of the aperture so that the diameter of the jet diminishes and becomes less than that of the aperture.

If a jet of water strikes against a body it exerts a force on the body. If the body moves, owing to the action of this force, the

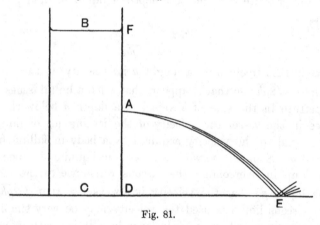

Fig. 81.

water does work on it and so imparts to the body some of its kinetic energy. The greatest possible amount of work is done on the body by the water when the water is reduced to rest. The force exerted by a jet of water is made use of in a form of water wheel called Pelton's wheel, shown in Fig. 82. The water issuing at A impinges on a series of buckets forming the wheel which are designed so as to reduce the water nearly to rest when the wheel is rotating at the proper speed. If the pressure of the water supply to the wheel is due to a depth h feet and n pounds of water enter the wheel per minute, then the theoretically possible horse-power of the wheel is equal to $\dfrac{nh}{33,000}$ since one horse-power is equal to 33,000 foot-pounds per minute. In practice the power obtained is always rather less than this.

If the lower end of a vertical glass tube of small diameter is dipped into a liquid like water which wets the glass, it is found that the water rises up the tube above the level of the free surface outside, to a height which is inversely proportional to the diameter of the tube.

Surface Tension.

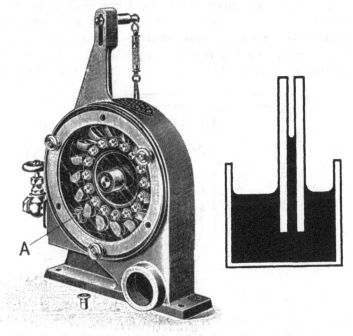

Fig. 82.   Fig. 83.

Fig. 83 shows a vertical section of such a tube dipping into a small vessel of water. Close observation shows that the surface of the water in the tube is not flat but is nearly hemispherical and concave upwards. Also round the walls of a glass vessel containing water the free surface curves up so that it meets the glass at a very small angle. Very small bubbles of air in water are nearly spherical in shape, and small drops of water resting on a surface, such as lamp black, which water does not wet are also nearly spherical.

These facts suggest that the surface of a liquid behaves as if it were covered by an elastic skin which tends to contract. Such phenomena are said to be due to surface tension. The surface of any liquid is in a state of tension, and tends to contract so as to make the surface as small as possible. Some liquids, like soap solution, can exist in thin films which tend to contract. Soap bubbles are a beautiful example of this. The force per unit length which the surface of a liquid can exert is taken as a measure of its surface tension. A water surface in air tends to contract with a force of about 75 dynes per centimetre. Owing to the tension in a thin film of soap solution there is an excess of pressure in a soap bubble over that outside. Fig. 84 shows an

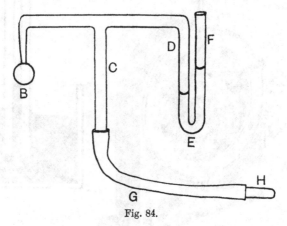

Fig. 84.

apparatus with which this excess of pressure can be measured. It consists of a T-shaped glass tube bent as shown, so that the part *DEF* forms a small water manometer. A piece of india-rubber tubing is fastened to the branch *C* and its end is closed by a glass rod *H*. The narrow neck at *B*, which should be not more than 0·2 cm. wide, is dipped into a little soap solution, and then by slightly pushing in the rod *H* a small soap bubble a few mms. in diameter is blown at *B*. The difference between the water levels in the manometer and also the diameter of the bubble are measured. It is found that the pressure in the bubble is inversely proportional to its diameter.

Let *ABC* (Fig. 85) be a soap bubble and let *AB* be an

imaginary plane passing through its centre. Let its radius be r, and let the difference between the pressure inside the bubble and that outside be denoted by p. If the surface tension of the soap solution is T then each cm. of the thin film, which has two surfaces, exerts a force $2T$. The two halves of the bubble on either side of the plane AB are therefore pulled together by a force $2T \times 2\pi r$. This force is balanced by the pressure p acting on the area πr^2 so that

$$p\pi r^2 = 4T\pi r \quad \text{or} \quad p = \frac{4T}{r}.$$

If we get p in dynes per sq. cm. and r in cms. this formula gives T in dynes per cm. For soap solution T is about 25 dynes per cm.

Thus if $r = 0.25$ cm. we get $p = \dfrac{4 \times 25}{0.25} = 400$ dynes per sq. cm. This would give a difference of level in the water manometer of $\dfrac{400}{980} = 0.408$ cm.

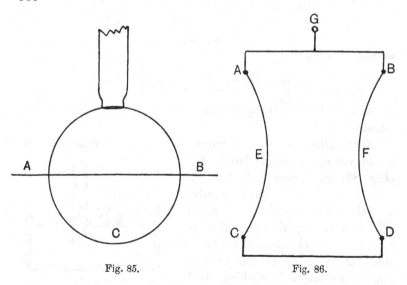

Fig. 85.   Fig. 86.

Another way of getting T for a soap film is with the apparatus shown in Fig. 86. This consists of two pieces of thin aluminium wire AB and CD bent as shown and connected by two fine silk threads AEC and BFD of equal length. If this arrangement is held by the ring G, dipped into soap solution and then withdrawn, a soap film is formed over the area $ABCD$. This film pulls in the threads so that they form arcs of circles as shown. Let $AB = CD = b$. Also let $AC = BD = h$ and let $EF = d$.

Let the mass of the wire CD be m grams. The wire CD is supported by the surface tension of the film and by the tensions in the two threads AEC and BFD. The weight of the soap film can be neglected, so that considering the forces across a horizontal line at EF we see that

$$mg = 2t + 2Td,$$

where t is the tension in each thread. If we consider a complete circle of thread surrounded by a soap film, it is easy to see that the tension in the thread is given by the equation

$$t = 2rT,$$

where r is the radius of the circle. It can also easily be shown from the geometry of the circle that the radius of the arcs AEC and BFD is given by the equation

$$r = \frac{h^2 + (b - d)^2}{4(b - d)},$$

so that we have

$$mg = 2T \left\{ \frac{h^2 + (b - d)^2}{2(b - d)} + d \right\}.$$

By means of this formula T can be found if h, b, d and m are measured.

The height h to which a liquid rises in a narrow tube which it wets can easily be calculated. At A (Fig. 87) the pressure is less than atmospheric by an amount $h\rho g$, where h is the height of A above the free surface of the liquid at B. Also since the surface of the liquid in the tube is spherical, the pressure due to its surface tension is equal to $2T/r$ where r is the radius of the tube. Hence $h\rho g = 2T/r$ or $h = 2T/\rho gr$. The pressure due to the surface tension is only half that in a soap bubble of radius r, because the soap film has two surfaces. The value of h can also be obtained by equating

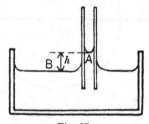

Fig. 87.

the upward force due to surface tension round the circumference of the tube to the downward force due to the pressure $h\rho g$ acting over the cross section of the tube. Thus

$$2\pi r T = h\rho g\pi r^2$$

or $\qquad\qquad h = 2T/\rho gr$ as before.

Since each unit length of a liquid surface exerts a force T in a direction perpendicular to the unit length and parallel to the surface it follows that a length l exerts a force lT. If a boundary of the surface of length l is moved in the plane of the surface parallel to itself a distance d, the work done against the force lT is equal to dlT. But dl is then equal to the increase in the area of the liquid surface, so that we see that to increase the area of the surface of a liquid requires an amount of work to be done equal to T units of work per unit area. If T is expressed in dynes per cm., then the work required is T ergs per sq. cm. The surface tension of a liquid can be explained by supposing that the molecules of the liquid attract each other when close together with a great force. In the interior of the liquid the molecules are free to move about because the forces on any molecule due to the surrounding molecules balance each other on the average. But when a molecule is close to the surface of the liquid there is a force on it in a direction perpendicular to the surface towards the interior of the liquid. This force tends to make molecules move from the surface into the interior of the liquid so that the surface tends to become as small as possible, that is, it tends to contract like a stretched sheet of india-rubber.

When a solid body is put in contact with a liquid, in many cases the solid dissolves in the liquid forming a
Diffusion.   solution. For example, sugar dissolves in water. If the bottom of a vessel is covered with a layer of sugar and then water is poured over the sugar and left undisturbed, it is found that the sugar gradually penetrates into the water so that after a long time, say a year or two, the sugar becomes uniformly distributed throughout the volume of the water.

If a solution of potassium permanganate in water is put at the bottom of a tall glass jar and water very slowly and gently poured on to it, it is possible to fill the jar with water without disturbing the solution. The potassium permanganate has an intense purple colour, and it is found that the colour gradually rises through the water until after a long time the whole of the water in the jar becomes equally coloured.

The molecules of a liquid are supposed to be in motion relatively to each other, so that any particular molecule moves about in the liquid and so in the course of a very long time will have been as long in one part of the vessel containing the liquid as in any other part. The molecules of substances, like sugar or potassium permanganate, also move about in the liquid in the same way, so that they gradually distribute themselves equally throughout the volume of the liquid. This process by which one substance penetrates into another is called diffusion. In liquids it is a very slow process.

By stirring up the liquid the uniform distribution of a substance dissolved in part of it can be brought about in a few seconds.

PART II

HEAT

CHAPTER I

TEMPERATURE

THE words hot, warm, cold, cool are used to indicate different
states of bodies with which everyone is more or
less familiar. If a cold body is put near a fire
it gradually becomes warm and then hot, and if put in the fire
it may become red hot or even white hot. Thus a body can pass
from a state in which it is cold to a state in which it is white hot
by a continuous process, so that there is an indefinitely large
number of possible states intermediate between a cold state and
a hot state. The temperature of a body is a quantity which
indicates how hot or how cold the body is. In order to deal
with temperatures scientifically it is necessary to adopt some
measurable property of a body as a measure of its temperature,
and to define a unit of temperature.

Temperature.

It is found that bodies expand or get larger when they are
made hotter and that liquids usually expand more
than solids. For example if some mercury is put
in a bulb blown on the end of a glass tube of
small bore, so that the mercury completely fills the bulb and a part
of the tube, then it is found that as this arrangement gets hotter
the mercury increases in volume so that it occupies a greater part
of the tube. Such a bulb and tube containing mercury is called
a mercury-in-glass thermometer. The position of the end of the
mercury column in the tube is taken as an indication of the temperature of the thermometer. We may suppose a scale of equal

The Mercury-
in-Glass
Thermometer.

lengths marked on the tube so that the position of the mercury
can be read off on the scale. If the bulb is put in contact with a
hot body, then the mercury rises to a certain point, and if it is then
put in contact with a cold body it falls. If the hot body and the
cold body are put in contact it is found that the hot body gets
colder and the cold body gets hotter, and after a time the mercury
in the thermometer is found to stand at the same level when the
bulb is in contact with either body. If a number of bodies, some
hot and some cold, are put in a wooden box lined with copper and
left there for a time, then the thermometer will give the same
indication when in contact with any of the bodies in the box.
We conclude then that bodies in contact with each other, when
protected from any disturbing cause which might tend to make
some of them hotter than the others, all take up equal temperatures. When a thermometer bulb is immersed in a liquid its
temperature soon becomes the same as that of the liquid, and
when the bulb is put into a hole bored in a solid body it takes up
the temperature of the body. A thermometer therefore serves to
indicate not only its own temperature but also the temperature of
any body with which it is in sufficiently intimate contact.

If a thermometer is put into a mixture of crushed ice and
pure water, it is found that the mercury stands at a certain
level which remains constant so long as there is plenty of ice
mixed with the water. The indication of the thermometer does
not change if the mixture is taken from a cold place to a warm
place, although this causes the ice to begin to melt. If the same
thermometer is put into mixtures of ice and water made up in
different places and at different times, it is found always to
indicate the same temperature. We conclude therefore that the
temperature of a mixture of ice and water is a constant definite
temperature. This temperature is called the melting point of
ice. In the same way if a thermometer is put into the steam
rising from boiling water at a pressure of 76 cms. of mercury, then
it always indicates the same temperature. This temperature is
called the boiling point of water at 76 cms. of mercury pressure.

In order to establish a scale of temperature it is necessary to
devise some way of graduating thermometers so that they will
all agree in their indications. The plan adopted is to make use

of the definite temperatures given by mixtures of ice and water and by steam from boiling water at 76 cms. of mercury pressure. The thermometers should all be made of the same kind of glass and be filled with pure mercury. The cross section of the tube should be constant throughout its length. The position of the mercury is marked on the tube first when the thermometer is in a mixture of pure ice and water and then when it is in steam from boiling water at 76 cms. of mercury pressure.

The distance between these two marks, which are called the fixed points of the thermometer, is then divided into a number of equal parts which are said to correspond to degrees of temperature. The centigrade scale of temperature is obtained by dividing the distance between the fixed points into 100 equal parts. The freezing point of water is called 0°C. and the boiling point 100°C. By continuing the scale of equal divisions below the freezing point and above the boiling point we can extend the scale to higher and lower temperatures.

The Fahrenheit scale of temperature is obtained by dividing the distance between the fixed points into 180 equal parts. The boiling point is called 212°F. and the freezing point 32°F., so that the zero on this scale of temperature is 32 degrees below the freezing point of water. The centigrade scale is always used in scientific work. Fig. 1 shows two thermometers, one graduated on the centigrade scale and the other on the Fahrenheit scale. The relation between the two scales is indicated by this figure. Temperatures measured with different thermometers, graduated in the way described, agree very nearly with each other. For example, if two such thermometers are immersed in a vessel of water which is kept stirred, they will both indicate very nearly the same temperature. It is found, however, that there are small differences between the indications of the different thermometers when they are really at equal temperatures. This is due to the tubes not being of exactly uniform cross section and to differences between the expansions of the glass bulbs. In accurate work it is necessary to eliminate these small differences. This is done by having all thermometers intended for use in accurate work compared with a standard thermometer. The Bureau of Standards at Washington, U.S.A. and similar institutions in other countries

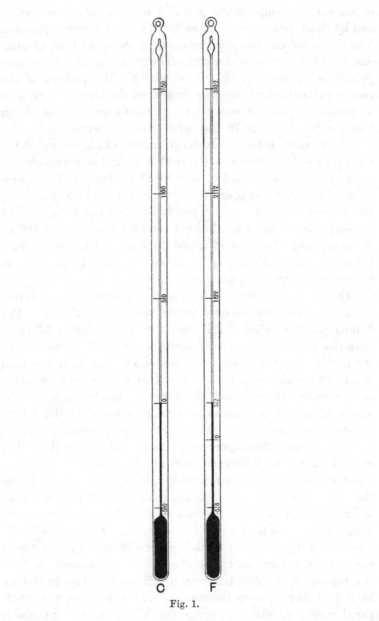

Fig. 1.

undertake such comparisons. A thermometer can be sent to the Bureau of Standards, and it is there compared with a standard thermometer and a table giving the differences between its indications and those of the standard is made and sent back with it. In this way the small differences between different thermometers can be eliminated. We shall see later that it is possible to define a scale of temperature which does not depend on the properties of particular substances like glass and mercury. This scale is called the absolute scale of temperature. It is found that the scale of temperature given by a good mercury-in-glass thermometer agrees very nearly with the absolute scale between 0°C. and 100°C.

Mercury freezes at −40°C., so that a mercury-in-glass thermometer cannot be used below this temperature. Thermometers containing alcohol instead of mercury can be used down to about −112°C., and thermometers containing pentane down to −200°C. Mercury-in-glass thermometers cannot be used much above 300°C., but mercury in fused quartz thermometers can be used up to 450°C. The tubes of thermometers for use at high temperatures are filled with nitrogen gas, the pressure of which prevents the mercury from boiling. Temperatures above 300°C. are usually not measured with mercury thermometers, but by other methods some of which will be described in later chapters.

The following are some important temperatures on the centigrade scale:

Lowest possible temperature	−273
Boiling point of liquid hydrogen	−253
Boiling point of liquid oxygen	−183
Melting point of mercury	−40
Melting point of ice	0
Boiling point of water	100
Boiling point of sulphur	444·5
Melting point of gold	1061
Melting point of platinum	1780

REFERENCES

Theory of Heat, J. Clerk-Maxwell.
Heat, Poynting and J. J. Thomson.

CHAPTER II

THE EXPANSION OF SOLID BODIES WITH RISE OF TEMPERATURE

IT is found that solid bodies get larger when they are made hotter. For example an iron rod, which is 100 cms. long when in a mixture of ice and water, becomes about 100·12 cms. long when surrounded by steam from boiling water. It is found that the increase of length per unit length is approximately proportional to the rise in temperature. Thus if l_1 denotes the length of a body at $t_1°$C. and l_2 the length at $t_2°$C., then

$$l_2 - l_1 = \alpha l_1 (t_2 - t_1)$$

or $$l_2 = l_1 \{1 + \alpha (t_2 - t_1)\},$$

where α is a constant. α is equal to the increase in length of unit length for one degree rise in temperature; it is called the coefficient of linear expansion. The expansion of rods can be measured with the apparatus shown in Fig. 2. A rod AB about

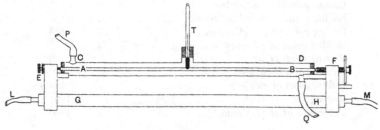

Fig. 2.

50 cms. long and 0·5 cm. in diameter is contained in a brass tube CD about 2 cms. in diameter and a few millimetres shorter than the rod. The ends of the tube are closed by short corks, through which the ends of the rod project. The temperature of the rod is

measured with a thermometer T. The tube CD rests on two V-shaped supports carried by a metal frame $EGHF$. At E one end of the rod is in contact with a screw, and at F there is a micrometer screw which can be adjusted so that the distance between the ends of the two screws is just equal to the length of the rod. The base of the frame $EGHF$ consists of a tube through which cold water is passed by means of the india-rubber tubes L and M. In this way the frame is kept at a constant temperature. Cold water is first passed through the tube CD, and the micrometer adjusted and its reading noted. The micrometer is then screwed out about 0·5 cm. and steam passed through the tube CD by means of the rubber tubes P and Q. The rod is thus heated to about 100°C. and its length increases. The micrometer is screwed in till it just holds the rod so that it cannot be shaken in a direction parallel to its length. The increase in length of the rod is got from the change in the micrometer screw reading and the temperatures are given by the thermometer. The length at different temperatures can be obtained by passing hot water instead of steam through the tube CD. With such an apparatus the change in length can be measured to about 0·005 millimetre.

A more exact method of finding the change of length of a rod when heated is to compare its length with that of another rod of nearly equal length kept at a constant temperature. The rods may be flat bars about 105 cms. long, each having two narrow transverse lines ruled on them 100 cms. apart. One of the rods is kept in a tank filled with a mixture of ice and water and the other in a tank of water which can be heated. The temperature of the water is got with several thermometers, the bulbs of which are put close to the bar at equal distances along its length. The water in the tank is stirred all the time so as to keep its temperature as uniform as possible.

The lengths of the two rods are compared by means of two microscopes firmly fixed one metre apart to rigid supports. One of the bars in its tank is placed beneath these microscopes, and its position is adjusted until the transverse lines on it can be seen in the microscopes. The eyepiece of each microscope contains a cross hair which can be moved sideways by means of a micrometer screw. These cross hairs are adjusted until they appear to coincide with the transverse lines on the bar as seen through the microscopes. The other bar in its tank is then brought beneath the microscopes and its position adjusted until one of the transverse lines on it appears to coincide with one of the cross hairs. If the other line also appears to coincide with the other cross hair, then the distance between the lines on the second bar is equal to that on the first bar. If it does not appear to coincide, then the cross hair is moved until it does appear to coincide and the distance it has to be moved through is measured with the micrometer screw. If a standard scale divided into millimetres is observed with the microscopes then it is easy to find how far their cross hairs have to be moved to make them appear to go from one scale division to the next one. The difference between the lengths of the two bars, corresponding to the distance the cross hair had to be moved, can then be calculated. In this way the change in the length of the bar in the tank of water when the temperature of the water is changed can be accurately determined. The change in the length can be found in this way to less than 0·001 millimetre. The length of the bar in the ice and water can be found by comparing it with a standard metre also kept in a tank of ice and water or a standard metre may be used instead of this bar. The apparatus described is called a comparator and is the most accurate instrument known for finding the difference between two nearly equal lengths. It is always used for comparing the lengths of standards of length.

The following table gives the values of the coefficients of linear expansion for bars made of several substances:

Brass	...	...	...	...	...	0·000019
Iron	...	...	...	...	...	0·000012
Glass	...	...	...	...	...	0·000009
Copper ...	...	...	...	...	0·000017
Platinum	...	...	...	...	0·000009
Fused quartz ...	...	...	...	0·0000007
Nickel steel 36 per cent. Ni	...	0·0000009

A brass bar 100 cms. long at 0°C. increases in length by 1·9 mms. when heated to 100°C. while a bar of fused quartz of the same length increases by only 0·07 mm. It is found that different samples of the same solid substance do not all expand exactly equally, so that when it is necessary to know the coefficient of expansion of a particular sample exactly it has to be measured. The expansion of solids with rise of temperature is not exactly uniform, so that the formula

$$l_2 = l_1 \{1 + \alpha (t_2 - t_1)\}$$

is not exactly true. When $t_2 - t_1$ is not more than, say, 100°C., this formula is sufficiently nearly true for most purposes. The value of α is not quite independent of the temperature.

A more exact formula is

$$l_t = l_0 (1 + \alpha t + \beta t^2),$$

where l_t denotes the length at t°C., l_0 that at 0°C., and α and β are constants. For platinum between 0°C. and 1000°C.

$$\alpha = 0\cdot00000887 \quad \text{and} \quad \beta = 0\cdot000000001324.$$

Although the expansion of solids with rise of temperature is small, it is nevertheless of considerable practical importance in engineering and other branches of applied science. Large steel structures like bridges have to be designed so that they can expand and contract freely without damage, and the rails of car lines and railways have to be laid with short gaps between their ends to allow room for expansion.

The temperature out of doors may vary from say -30°C. to $+40$°C., a range of 70°C. The change of length of a steel structure 1000 feet long when its temperature changes 70°C. is about 10 inches. It may therefore be necessary to allow for possible changes in the lengths of steel or iron structures of about one inch per 100 feet.

The contraction of bodies on cooling is sometimes made use of for fixing things firmly together. For example, the iron tires of wheels are made a little smaller than the wheel and then slipped on while hot. In the manufacture of glass articles it is necessary to cool them very slowly, because otherwise some parts cool quicker than others, so that the articles are liable to be broken by the unequal contraction. Articles made of fused quartz can be heated and cooled in any manner without breaking because the expansion of fused quartz is so small.

The changes in the length of the pendulum of a clock, with changes of temperature, cause its time of vibration to alter so that the rate of the clock varies with the temperature. Accurate clocks are therefore provided with pendulums constructed so that their lengths remain constant in spite of temperature changes. Such pendulums are called compensated pendulums. One form of compensated pendulum has a bob consisting of a glass vessel containing mercury. When the temperature rises the pendulum gets longer, but at the same time the mercury expands and so its level in the vessel rises. Since mercury expands more than solids like brass and iron it is possible to design the pendulum so that the rise in the level of the mercury compensates for the increase in the length of the pendulum rod. If the pendulum rod is made of fused quartz or 36 % nickel steel its expansion is so small that it can be neglected except in the case of clocks intended for very exact measurements. Wood has a rather small coefficient of expansion, so it is often used for making the rods of clock pendulums. If the rod is made of a substance like wood which has a small coefficient of expansion, then the pendulum can be compensated by making the bob of some material like brass which has a larger coefficient. The rod is passed through a hole in the bob, and the bob is supported by a nut screwed on the end of the rod below the bob (Fig. 3). When the temperature rises the rod gets longer, but at the same time the bob expands so that its centre of gravity rises to a greater height above the nut on the rod. By making the bob the proper height the two effects can be made to compensate each other.

The time of vibration of the balance wheel of a watch is also affected by changes of temperature. As the temperature rises the wheel gets larger and also the restoring couple due to the spiral hair spring gets smaller. Both these effects make the time of vibration longer. The effect due to the change in the elasticity of the hair spring is much the greater of the two. These effects are compensated by making the rim of the balance wheel in two separate halves, each supported by a spoke at one end only (Fig. 4). Each half of the rim is made of a strip of brass, which is on the outside, with a strip of steel attached to it on the inside. When the temperature rises the brass expands more than the steel, so that the curvature of the rim is increased and the free ends of each half move in towards the axle. This diminishes the time of vibration, and by properly designing the wheel this effect can be made to compensate the increase in the time due to the other two effects.

If two equal bars, one of brass and one of steel, are brazed together so as to form a compound bar, then if the bar is straight when cold it becomes curved on heating. The brass side becomes convex and the steel side concave.

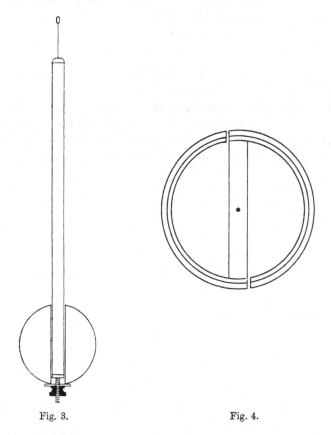

Fig. 3.   Fig. 4.

When a body is heated it expands in all directions, so that its volume gets greater. Consider a rectangular block having sides of length a_0, b_0, c_0 at $0°$ C. Its volume at $0°$ C. is therefore equal to $a_0 b_0 c_0$. If its temperature is raised to $t°$ C. the lengths of its sides are given by the equations

$$a_t = a_0 (1 + \alpha t),$$
$$b_t = b_0 (1 + \alpha t),$$
$$c_t = c_0 (1 + \alpha t),$$

where α is the coefficient of linear expansion of the block. Its volume at $t°C$. is therefore equal to

$$a_0 b_0 c_0 (1 + \alpha t)^3.$$

Since α is a small quantity this is nearly equal to $a_0 b_0 c_0 (1 + 3\alpha t)$. Let $a_t b_t c_t = v_t$ and $a_0 b_0 c_0 = v_0$, so that

$$v_t = v_0 (1 + 3\alpha t)$$

or $\qquad v_t = v_0 (1 + \beta t)$, where $\beta = 3\alpha.$

β is called the coefficient of cubical expansion and is equal to three times the coefficient of linear expansion.

If a brass sphere is made so that it can just pass through a brass ring when both are cold, then on heating the sphere it is found to be too large to go through the ring. If the ring is then heated it becomes big enough to let through the hot sphere.

CHAPTER III

THE EXPANSION OF LIQUIDS WITH RISE OF TEMPERATURE

It is found that liquids expand when they get hotter. The fact that mercury expands more than glass is evident from the rise of the mercury in a thermometer tube when the temperature of the thermometer gets higher.

Let the volume of a thermometer bulb and tube at $t°$C., up to the freezing point mark, be denoted by b_t, and let $b_t = b_0 (1 + \beta t)$, where b_0 is the volume at $0°$C. and β the coefficient of cubical expansion of the glass. Let the volume of mercury at $t°$C. be denoted by V_t. Then since at $0°$C. the mercury just fills the bulb and tube up to the freezing point mark we have $V_0 = b_0$. Let then $V_t = b_0 (1 + \gamma t)$, where γ is the coefficient of cubical expansion of the mercury. The volume of the mercury in the tube beyond the freezing point mark, when the thermometer is at $t°$C., is then given by

$$V_t - b_t = b_0 \{1 + \gamma t - 1 - \beta t\} = b_0 t (\gamma - \beta).$$

If β were equal to γ then $V_t - b_t$ would be zero, so that the mercury would not rise in the tube at all. The quantity $\gamma - \beta$ is called the coefficient of apparent expansion of mercury in glass because the amount of mercury which passes the freezing point mark is the same as it would be if the glass did not expand at all and the mercury had a coefficient of cubical expansion equal to $\gamma - \beta$.

The expansion of liquids can be easily measured by means of an apparatus similar to a thermometer bulb and tube called a dilatometer. One form of dilatometer is shown in Fig. 5. It consists simply of a glass bulb of about 5 c.c. capacity blown on a tube about 15 cms. long and about 1 mm. internal diameter. The tube is graduated into millimetres. The bulb and tube can be easily filled with any liquid by means of a glass funnel with a long neck only 0·5 mm. in diameter made by drawing out a glass tube in a blowpipe flame. Such a funnel is shown in Fig. 6. To empty the dilatometer the funnel is connected to a vacuum pump and the liquid drawn out by means of it. The volume of the dilatometer

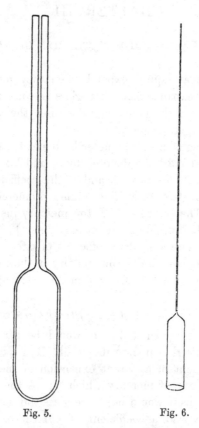

Fig. 5.   Fig. 6.

at 0°C. up to any division on the tube can be found by filling it with mercury at 0°C. and getting the weight of the mercury. One c.c. of mercury at 0°C. weighs 13·5955 grams. If a liquid is put in the dilatometer its apparent volume at any temperature is equal to the known volume at 0°C. of the portion of the dilatometer which it occupies. The true volume can be calculated from the volume at 0°C. by means of the equation $V_t = V_0 (1 + \beta t)$, where β is the coefficient of cubical expansion of the glass. If the dilatometer is made of fused quartz instead of glass its expansion is then so small that it can be neglected except when very great accuracy is required. It is found that the volume of most liquids does not increase very uniformly as the temperature rises. If V_1 denotes the volume of a liquid at $t_1°$C. and V_2 the volume at $t_2°$C., then if $t_2 - t_1$ is only a small number of degrees, say 20 or 30, it is usually nearly true that

$$V_2 - V_1 = \beta V_1 (t_2 - t_1),$$

where β is a constant. If it is desired to represent the volume of a liquid over a large range of temperature, then the formula

$$V_t = V_0 (1 + c_1 t + c_2 t^2 + c_3 t^3)$$

may be used where c_1, c_2 and c_3 are constants.

The following table gives some values of these constants :

Liquid		c_1	c_2	c_3
Alcohol	− 39 to 27°C.	0·001033	0·0$_5$145	
,,	27 to 46°C.	0·001012	0·0$_5$220	
Mercury	24 to 299°C.	0·0$_3$18116	0·0$_7$1155	0·0$_{10}$21187
Ether	− 15 to 38°C.	0·0015132	0·0$_5$23592	0·0$_7$400512
Benzene	11 to 81°C.	0·0011763	0·0$_5$12776	0·0$_8$80648

Here, for example, 0·0$_5$145 means 0·00000145. The dilatometer method requires the expansion of the dilatometer to be known. The best way to find it is to use the dilatometer to measure the expansion of a liquid the expansion of which has been found independently.

The expansion of mercury has been found by several observers by means of a method invented by Dulong and Petit which is independent of the expansion of the containing vessel. Fig. 7 shows a simple form of apparatus for finding the expansion of mercury by this method. It consists of a glass tube $ABCD$, of which the parts AB and CD are vertical and the part BC horizontal. The vertical parts are enclosed in two wide tubes as shown, which are separated by a wooden board. Steam is passed through one of the wide tubes and cold water through the other. The tube $ABCD$ is nearly filled with mercury which stands at a higher level on the hot side than on the cold side. Let t_1 denote the temperature of the mercury in AB, t_2 that of the mercury in CD, and let h_1 be the height of the column of mercury in AB and h_2 that of the column of mercury in CD. Then since the two columns of mercury balance each other the pressures due to them must be equal, so that $h_1\rho_1 g = h_2\rho_2 g$, where ρ_1 is the density of the mercury in AB and ρ_2 that of the mercury in CD. We have therefore $h_1/v_1 = h_2/v_2$, where v_1 and v_2 denote the volumes of one gram of mercury at the temperatures t_1 and t_2. Also

$$v_1 = v_0\,(1 + \gamma t_1) \text{ and } v_2 = v_0\,(1 + \gamma t_2),$$

where γ is the coefficient of cubical expansion of the mercury. Hence

$$\frac{h_2}{h_1} = \frac{1 + \gamma t_2}{1 + \gamma t_1}.$$

This equation enables γ to be calculated from h_2, h_1, t_2 and t_1.

With an apparatus similar in principle to that just described Regnault found the density of mercury at temperatures from $0°$C. to $360°$C. His results are represented by the formula

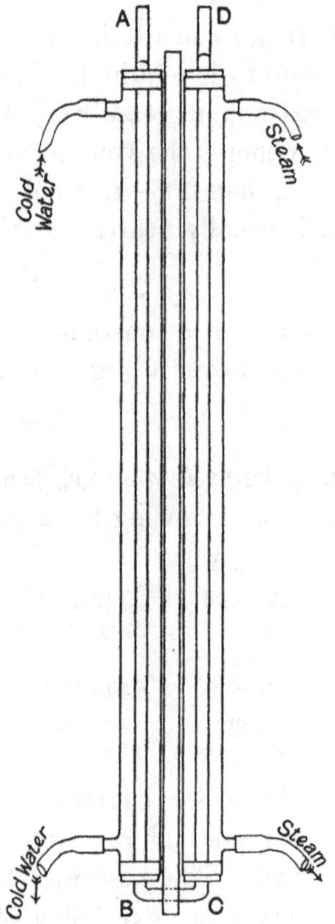

$$\frac{V_t}{V_0} = \frac{\rho_0}{\rho_t} = 1 + c_1 t + c_2 t^2 + c_3 t^3,$$

Fig. 7.

where the c's have the values given in the table above. The following table gives the volume of 100 grams of mercury at several temperatures:

Temperature	Volume	Differences
$0°$ C.	7·35540	
50	7·42229	0·06689
100	7·48939	0·06710
150	7·55688	0·06749
200	7·62495	0·06807
250	7·69381	0·06886
300	7·76364	0·06983

The third column gives the differences between the numbers in the second column. We see that the increase in volume for a rise of 50°C. is not constant but gradually gets greater as the temperature rises. The increase from 0°C. to 50°C. is nearly equal to that from 50°C. to 100°C., so that over the range of temperature 0°C. to 100°C. the simple equation $V_t = V_0(1 + \beta t)$ represents the volume of mercury very well. β is equal to about 0·0001822.

The density of a liquid at different temperatures can be found by suspending a body in it by a fine wire from one pan of a balance and getting its apparent weight at different temperatures. If V_t denotes the volume of the body and ρ_t the density of the liquid, then the apparent weight is less than the actual weight by $V_t \rho_t$ grams. If the coefficient of cubical expansion of the body is β, then $V_t = V_0(1 + \beta t)$, so that if W_t denotes the apparent loss of weight we have

$$W_t = \rho_t V_0 (1 + \beta t)$$

or

$$\rho_t = W_t / V_0 (1 + \beta t).$$

If the body is made of fused quartz β is equal to only 0·0₆7, and so can be neglected except in very exact work. We have $\rho_0 = W_0/V_0$, so that

$$\frac{\rho_t}{\rho_0} = \frac{W_t}{W_0 (1 + \beta t)}.$$

This method has been used to find the density of water at different temperatures. The following table gives some values of the density of water in grams per cubic centimetre and also of the volume of one gram :

Temperature	Density	Volume of 1 gram
0°C.	0·999868	1·000132
2	0·999968	1·000032
4	1·000000	1·000000
6	0·999968	1·000032
8	0·999876	1·000124
10	0·999727	1·000273
20	0·998230	1·001773
30	0·995673	1·004346
50	0·98807	1·01207
70	0·97781	1·02270
100	0·95838	1·04343

It will be seen that from 0°C. to 10°C. the density is nearly constant and equal to one, but it is slightly less at 0°C. than at 4°C., and slightly less at 8°C. than at 4°C. Thus the density of water has a maximum value at 4°C. Above 8°C. the density diminishes as the temperature rises, but not at a uniform rate. The rate of diminution increases as the temperature rises.

If a tall jar of water is surrounded by crushed ice contained in a larger jar, then the water will slowly cool down. Let the temperature of the water at several depths below its free surface be observed by means of a thermometer supported in it. If the water is not disturbed it will be found that at first the water at the bottom is colder than that at the top of the jar. When the water at the bottom gets down to about 4°C., however, it stops getting colder, and the water at the top gets colder than that at the bottom. The water at the top eventually gets down to 0°C. and this temperature gradually spreads from the top down to the bottom of the jar. These facts can be easily explained. When all the water is above 4°C. the coldest water is also the heaviest and so sinks to the bottom, but when some of the water gets down to 4°C. then it is as dense as possible, and water below 4°C. floats above it. The coldest water therefore stays at the top when the temperature gets below 4°C.

Lakes and rivers in cold weather cool down like the jar of water just considered. When all the water has got down to 4°C. water colder than this stays at the top, so that ice forms first at the top. The water below the surface after it has got down to 4°C. is not further cooled by streams of colder water sinking down, so that lakes and rivers do not usually cool below 4°C., except near the surface, even in very severe winters.

The contraction when water at 0°C. is heated can be shown by means of a fused quartz dilatometer. If one is filled with water and slowly cooled the level of the water falls very slowly when the temperature approaches 4°C., and then it stops falling and eventually rises as the temperature falls below 4°C. down to 0°C. On warming the dilatometer the water first falls and then rises again.

CHAPTER IV

THE PROPERTIES OF GASES

GASES always completely fill the vessel containing them; so
that the volume of a gas is equal to
the volume of the vessel in which it is
enclosed. The gas exerts a pressure on
the walls of the vessel which depends
on the volume of the vessel, on the
temperature and on the mass and
nature of the gas in the vessel.

If the temperature of a quantity of
gas is kept constant then it is found
that the pressure varies very nearly
inversely as the volume. This law was
discovered by Boyle in 1662 and is
known as Boyle's law. It may be
verified with the apparatus shown in
Fig. 8. This consists of a U-shaped
glass tube $ABCD$ closed at A and open
at D. At the bottom a side tube is
joined on to it and an india-rubber
tube leads from this to a vessel E.
The vessel E, the rubber tube, and
part of the U tube, contain mercury.
The mercury encloses a volume of dry
air, or other gas, in the closed tube
AB. The tube AB is surrounded by
a wider tube EF, the lower end of
which is closed by a cork through which
AB passes. A current of water at a
known temperature can be passed
through EF so as to keep the gas in

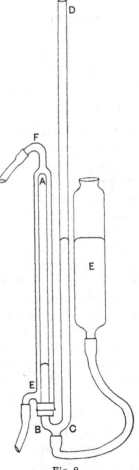

Fig. 8.

AB at a constant temperature. A millimetre scale is marked on
the tube AB or else one is fixed up alongside it. If the tube AB
is of uniform bore the length of the part of it which is full of gas
is proportional to the volume of the gas. The pressure of the gas
in AB is equal to the atmospheric pressure outside plus the
pressure due to the difference of level between the mercury in CD
and that in AB. By raising or lowering the vessel E the level of
the mercury in CD can be altered and so the pressure on the gas
in AB varied. The following table gives a series of results obtained
with such an apparatus at a temperature of $15°$ C.

(1) Length of AB containing air	(2) Height of mercury in CD above that in AB	(3) Pressure	(4) Product of numbers in columns (1) and (3)
20 cms.	2·0 cms.	77·0	1540
15 „	27·8 „	102·8	1542
10 „	79·0 „	154·0	1540

The height of the barometer was 75·0 cms. of mercury. The
pressures in column (3) are obtained by adding the numbers in
column (2) to the barometer height. The third column contains
the products formed by multiplying these pressures by the corresponding lengths of the gas in AB. These products are nearly
constant, which shows that the volume of the gas is inversely
as its pressure. The density of a gas is inversely proportional
to its volume, so that another way of stating Boyle's law is to
say that the density of a gas is nearly proportional to its pressure.
If a definite mass of a gas, say one gram, is admitted into an
empty vessel it fills the vessel and exerts a certain pressure on
the walls. If now another gram of the same gas is admitted
the density of the gas in the vessel is doubled and the pressure
is also doubled. Thus the second gram produces the same increase
of pressure that it would have produced if the first gram had not
been there. If a third gram of the gas is let in, the pressure
becomes three times that due to one gram, so that the third gram
produces the same increase of pressure as it would have produced
if the vessel had been empty. Thus it appears that the pressure
due to a gas is equal to the sum of the pressures each part of the
gas would exert if present by itself in the same vessel.

It is supposed that gases consist of molecules all in rapid

motion in different directions and that the total volume of all
the molecules in a gas at ordinary pressures is very small compared with the volume of the vessel containing the gas. Thus
the greater part of the volume of the vessel is really empty even
when the vessel is filled with a gas. If more gas is let in the
molecules composing it move about in the empty spaces between
the other molecules and so produce the same pressure on the walls
that they would have produced if the vessel had been empty.
Although the volume actually occupied by the molecules is small,
they are so numerous that very many thousands are contained
in every cubic millimetre of the volume of the vessel.

It is found that Boyle's law is not exactly true. The following
table gives some values of the product of the pressure p and
volume v for different gases. If Boyle's law were true these
products would be independent of the pressure.

Air at 0° C.

p in atmospheres	v	pv
1	1,000,000	1,000,000
100	9730	973,000
200	5050	1,010,000
400	3036	1,214,400
800	2168	1,734,400

Hydrogen at 0° C.

p in atmospheres	v	pv
1	1,000,000	1,000,000
200	5690	1,138,000
400	3207	1,282,800
800	1972	1,577,600

The pressure p in the table above is expressed in atmospheres.
One atmosphere is equal to about 10^6 dynes per sq. cm. or 15
pounds weight per sq. inch. We see that when gases like air
and hydrogen are exposed to enormous pressures the product
does not remain constant but varies considerably. At 800 atmospheres or about six tons weight per square inch the product pv
for air at 0° C. is 75 % greater than at one atmosphere. At small
pressures of not more than say 200 cms. of mercury Boyle's law
is very nearly true for air, oxygen, nitrogen, hydrogen, helium and

some other gases.   Other gases, including carbon dioxide, sulphur
dioxide, and ethylene, show considerable deviations from Boyle's
law even at moderate pressures.

Boyle's law is approximately true, at any constant temperature,
provided the pressure is not made too great.   If steam is passed
through the tube EF in the apparatus shown in Fig. 8 the gas in
the tube AB is heated to about 100° C. and Boyle's law can be
shown to be true at this temperature.   The following table gives
some results obtained with the same amount of air as was used in
the experiment previously described.   Temperature 100° C.   Height
of barometer 75 cms.

Length of AB containing air	Pressure	Product
25 cms.	79·8 cms.	1995
20 „	99·8 „	1996
15 „	133·0 „	1995

It appears that the product of the pressure p and the volume v
at 100° C. is constant, as it was at 15° C., but it is equal to
1995 at 100° C. and 1540 at 15° C   It is found that the product
pv increases uniformly with the temperature.   If p_t denotes the
pressure and v_t the volume of a definite mass of gas at $t°$ C. and
p_0 and v_0 the pressure and volume of the same mass at 0° C., then

$$p_t v_t = p_0 v_0 (1 + \alpha t),$$

where α is a constant.   If we substitute the values 1995 at 100°C.
and 1540 at 15°C., we get

$$\frac{1995}{1540} = \frac{1 + 100\alpha}{1 + 15\alpha},$$

which gives α equal to nearly $\frac{1}{273}$.   It is found that for gases
like air, oxygen, nitrogen, hydrogen, helium, argon, carbon mon-
oxide, etc. for which Boyle's law is nearly true at moderate
pressures, the constant α has nearly the same value $\frac{1}{273} = 0\cdot00366$.

The equation $p_t v_t = p_0 v_0 (1 + \alpha t)$ shows that if the volume is
kept constant, so that $v_t = v_0$, then $p_t = p_0 (1 + \alpha t)$, and if the
pressure is kept constant, so that $p_t = p_0$, then $v_t = v_0 (1 + \alpha t)$.
Boyle's law is not exactly obeyed and the equations $p_t = p_0 (1 + \alpha t)$
and $v_t = v_0 (1 + \alpha t)$ are also not quite exact.

The variation of the pressure of a gas with its temperature when its volume is kept nearly constant can be examined with the apparatus shown in Fig. 9. This consists of a glass or fused quartz bulb B, joined to a narrow tube CD, which is joined at F to a wider vertical tube FE. At E an india-rubber tube K connects FE to another glass tube GH. At F there is a fine point of glass or quartz which projects to the axis of the tube FE as shown. The tubes FE and GH are filled with mercury. By raising or lowering GH the level of the mercury in EF can be adjusted so

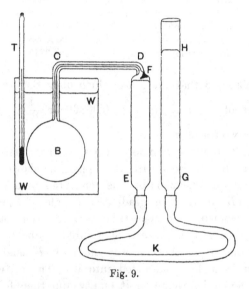

Fig. 9.

that the surface of the mercury just touches the sharp point at F. In this way the gas in B can be kept at a nearly constant volume. The bulb B is immersed in water or oil contained in a tank WW. The water is kept stirred and its temperature is measured with a thermometer T. The pressure of the gas in B is equal to the height of the barometer plus the difference of level between the mercury in GH and in EF. With this apparatus the pressure of the gas in B can be measured at a series of temperatures.

It is found that the pressure increases uniformly with the temperature. If the pressure is found when B is immersed in a mixture of ice and water, and also when it is immersed in steam from boiling water at 76 cms. of mercury pressure, the value of the constant α in the equation

$$p_{100} = p_0 \,(1 + 100\alpha)$$

can be found from these two pressures without using the thermometer T.

The following table gives some values of α found by Regnault. α is called the coefficient of pressure increase at constant volume.

Gas	α (volume constant)
Hydrogen	0·003667
Air	0·003665
Nitrogen	0·003668
Carbon monoxide	0·003667
Carbon dioxide	0·003688
Sulphur dioxide	0·003845
Cyanogen	0·003829

It will be observed that the gases which obey Boyle's law closely have values of α nearly equal to 0·00366 or $\dfrac{1}{273}$. The others have larger values of α.

The variation of the volume of a gas when its pressure is kept constant can be observed with the apparatus shown in Fig. 10. A glass or fused quartz bulb V is connected by a narrow tube to a tube AB on which is graduated a scale of equal volumes. This tube is joined to an open tube DE. At the bottom a side tube C is connected by means of a rubber tube to a vessel F containing mercury. A three-way stopcock K enables gas to be removed from V or admitted into it. The capacity of the bulb V is found by weighing it empty and then full of water. It is filled with dry air or other gas by means of an air pump and the stopcock K. The bulb V is immersed in a mixture of ice and water and the amount of gas in it is adjusted so that the mercury then stands at the zero of the scale on AB and is at the same level in ED. The cock K is then closed. If now the bulb V is immersed in steam from boiling water the gas in it expands and forces down the mercury in AB. By lowering F the mercury can be kept at the same level in DE and AB. In this way the pressure of the gas is kept constant at atmospheric pressure and the volume which comes out of V into AB can be read off on the scale.

Let the volume of the bulb be V c.c. If it is made of fused quartz its expansion can be neglected. Let the volume of the gas which comes out into AB be v and let the temperature of AB

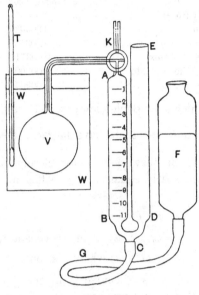

Fig. 10.

be $t°$ C. This temperature can be found by means of a thermometer put in contact with AB. Let α' denote the coefficient of expansion of the gas at constant pressure so that

$$V_t = V_0 (1 + \alpha't).$$

The volume v of gas at $t°$C. would become

$$v \frac{1 + 100\alpha'}{1 + t\alpha'} \text{ at } 100° \text{ C.}$$

Thus a volume V at $0°$ C. becomes

$$V + v \frac{1 + 100\alpha'}{1 + t\alpha'} \text{ at } 100° \text{ C.}$$

Hence　　　$$V + v \frac{1 + 100\alpha'}{1 + t\alpha'} = V (1 + 100\alpha),$$

or　　　$$v (1 + 100\alpha') = 100\alpha'V (1 + t\alpha').$$

By means of this equation α' can be calculated from the values found for v, V, and t. The following table gives some values of α' found by Regnault.

Gas				α' (pressure constant)
Hydrogen	...	...	...	0·003661
Air	...	...	...	0·003670
Carbon monoxide	...	...	0·003669
Carbon dioxide	...	...	0·003710
Sulphur dioxide	...	...	0·003903
Cyanogen	...	...	...	0·003877

The gases which obey Boyle's law closely have values of α' nearly equal to $\frac{1}{273}$ while the others have larger values.

For gases which obey Boyle's law closely the pressure coefficient α is very nearly equal to the coefficient of expansion α'. For these gases

$$p_t v_t = p_0 v_0 \, (1 + \alpha t),$$

so that $\qquad p_t = p_0 \, (1 + \alpha t)$ when v is constant,

and $\qquad v_t = v_0 \, (1 + \alpha t)$ when p is constant.

The density of a gas of course is inversely as its volume, so that in stating the density of a gas it is necessary to state its pressure and temperature. The density of gases is usually stated at a pressure of 76 cms. of mercury and a temperature of 0° C. The density of a gas can be found by weighing a large glass bulb first when exhausted by means of an air pump and then when filled with the gas at a known pressure and temperature. The difference between the two weights gives the mass of the gas, and this divided by the volume of the bulb is equal to the density. Let ρ denote the density at a pressure p cms. of mercury and temperature $t°$ C. The density ρ_0 at 0° C. and 76 cms. of mercury is then given by the equation

Density of Gases.

$$\frac{p}{\rho} = \frac{76}{\rho_0} \, (1 + \alpha t) \text{ where } \alpha = \frac{1}{273},$$

provided the gas is one of those which obeys Boyle's law closely at pressures between 76 and p. The following table gives the values of the ratios of the densities of several gases to the density of air at 76 cms. and 0° C. The pressure due to the 76 cms. of mercury depends on the value of g. The standard gas pressure adopted is the pressure due to 76 cms. of mercury in latitude 45° at sea level. The value of g in latitude 45° at sea level is

$$980 \cdot 632 \, \frac{\text{cms.}}{\text{sec.}^2}.$$

Gas				Density ÷ density of air
Air ...	...	...	...	1
Oxygen	...	...	...	1·10535
Hydrogen	...	...	...	0·06960
Nitrogen	...	...	...	0·96737

These values were found by Lord Rayleigh. One cubic centimetre of air at 0° C. and 76 cms. of mercury weighs 0·001293 gram. It is found that the densities of different gases at the same temperature and pressure are very nearly proportional to the molecular weights of the substances composing the gases. For example, the densities of oxygen, hydrogen and nitrogen are nearly proportional to 2×16, $2 \times 1 \cdot 008$ and 2×14. The molecular weight of a gas or vapour can therefore be found approximately by dividing its density by that of oxygen and multiplying by 32. For example, the density of helium at 0° C. and 76 cms. of mercury pressure is found to be 0·0001787 gram per c.c. and the density of oxygen is 0·0014292, so that the molecular weight of helium is

$$\frac{0 \cdot 0001787 \times 32}{0 \cdot 0014292} = 4 \cdot 001.$$

For a gas like air or hydrogen at constant volume let us **Gas Thermometers.** assume that $p_t = p_0 (1 + \alpha t)$ exactly, and use this equation to define a new scale of temperature. If the pressure p_0 when the vessel containing the gas is immersed in a mixture of ice and water and the pressure p_{100} when the vessel is immersed in steam from boiling water at 76 cms. pressure are found, then

$$p_{100} = p_0 (1 + 100\alpha),$$

or

$$\alpha = \frac{p_{100} - p_0}{100 p_0}.$$

Hence
$$p_t - p_0 = \frac{p_{100} - p_0}{100}\, t,$$

or
$$t = 100\, \frac{p_t - p_0}{p_{100} - p_0}.$$

By means of this equation the value of the temperature corresponding to any pressure can be calculated. The apparatus described above for measuring the pressure of a gas at constant volume can therefore be used as a thermometer to determine temperatures. It is then called a constant-volume gas thermometer. The scale of temperature given by such a thermometer is not exactly the same as the scale of temperature given by a mercury thermometer and different gases give slightly different scales of temperature. The scale also depends, but very slightly, on the initial pressure of the gas at 0° C. For accurate work the scale of temperature adopted as the standard is the scale given by a constant volume hydrogen gas thermometer in which the pressure of the hydrogen at 0° C. is equal to 100 cms. of mercury. The standard hydrogen thermometer of the Bureau International at Sèvres, France, gives the scale of temperature adopted as the standard scale in all civilised countries. Mercury thermometers are compared with this standard thermometer and a table of corrections made for each mercury thermometer giving the difference between the temperatures indicated by the mercury thermometer and the temperatures indicated by the standard hydrogen thermometer. In this way it is possible to measure temperatures anywhere on the scale of the standard hydrogen thermometer at Sèvres.

The differences between the scale of an ordinary mercury thermometer and the standard scale of temperatures are very small between 0° C. and 100° C., so that except in very exact work they can be neglected.

The following table shows the corrections that must be added to the readings of a mercury thermometer of hard glass made by Tonnelot to obtain the temperatures on the standard hydrogen scale:

Mercury thermometer	Corrections
0° C.	0
20	− 0·085
50	− 0·103
70	− 0·072
100	0

Thus if the mercury thermometer in question indicates 50° C. the temperature on the standard scale is 49·897° C.

The apparatus described above for measuring the increase of volume of a gas at constant pressure can also be used to measure temperatures. If v_{100} and v_0 are found, then the equation $v_t = v_0 (1 + \alpha t)$ gives

$$t = 100 \, \frac{v_t - v_0}{v_{100} - v_0}.$$

The scale given by a constant pressure gas thermometer is not exactly the same as the scale of temperature given by a constant volume thermometer containing the same gas.

In experiments with gases it is often necessary to remove as much as possible of the gas from a closed vessel so as to leave it nearly empty. If all the gas in a vessel could be removed so as to leave it completely empty, the empty space would be what is called a perfect vacuum. It is possible by means of modern air pumps to reduce the amount of gas in a vessel to a one-thousandth-millionth part of the amount present when the vessel is filled with the gas at atmospheric pressure. Even then the vessel contains many millions of molecules per cubic centimetre. A perfect vacuum cannot be obtained by any known process.

Air Pumps.

A very convenient air pump for many purposes is shown diagrammatically in Fig. 11. It was invented by Gaede and is known as Gaede's rotary box pump. AB is a circular cylindrical box of brass. Inside this a steel cylinder CD is mounted on an axle at O about which it can be made to rotate by means of a small electric motor, not shown in the figure. The cylinder CD has a slot in it, in which two plates E and F can slide. These plates fit closely the inside of the box AB and are pressed outwards by springs in the slot. Two tubes G and H lead into the box as shown, and the aperture at H is closed by a valve V. When the cylinder is rotating in the direction of the arrow the plates E and F push the gas round so that it is forced out through the valve V. A closed vessel connected to G is very quickly exhausted, so that the gas pressure in it is reduced to 0·01 mm. of mercury. If air or other gas is let in at G, it is blown out at H. This pump, therefore, can be used to pump gas into a vessel by connecting the vessel to H. While the pump is working,

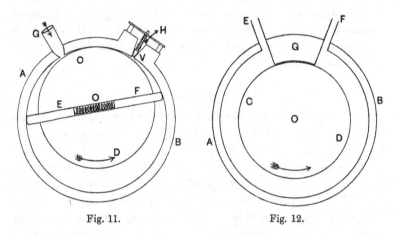

Fig. 11.   Fig. 12.

oil is continually supplied to the cylinder at its rotating shaft, and this oil is driven out through the valve along with the air. The oil supply is essential to the proper working of the pump.

To obtain very low pressures or very high vacua another form of pump also due to Gaede may be used. In Fig. 12, AB is a cylindrical box inside which a circular cylinder CD is mounted on an axle at O. Two tubes lead into the box at E and F and there is a projection at G on the inside of AB which nearly touches CD. If the cylinder CD is made to rotate at a very high speed, say 100 revolutions per second, the gas in the box is dragged round by the cylinder so that the gas pressure becomes higher at F than at E. It is found that the difference between the pressure at F and the pressure at E is nearly independent of the pressure at F. Thus if the pressure at F is 76 cms. of mercury that at E may be 74 cms. when it is connected to a closed vessel. If the pressure at F is reduced to 20 cms. then that at E becomes 18 cms. If the pressure at F is 3 cms. that at E is 1 cm. Suppose now F is connected to the rotary box pump and the pressure at F reduced to 0·01 mm. The pressure at E then becomes extremely small and may fall as low as 0·000001 mm. of mercury. This is probably the most perfect vacuum obtainable at the present time.

Many other forms of air pump are in use.

REFERENCE

The Experimental Study of Gases, Travers.

CHAPTER V

QUANTITY OF HEAT. SPECIFIC HEAT

IF a hot body is placed in contact with a cold body, the cold body gets hotter and the hot body colder. We say that this is due to heat passing from the hot to the cold body. Suppose for example that a piece of copper heated to 100°C. in the steam from boiling water is placed in a vessel of water at 0°C., and that the temperature of the water then rises to 4°C. The temperature of the copper at the same time falls from 100°C. to 4°C. A certain amount of heat has passed from the copper into the water. In order to deal with quantities of heat scientifically it is necessary to adopt a unit quantity of heat, and to devise methods of measuring quantities of heat in terms of the unit adopted. It is found that heat is a measurable quantity as we shall see presently. The unit quantity of heat generally used in scientific work is the amount of heat required to raise the temperature of one gram of water from 15°C. to 16°C. This amount of heat is called a *calorie*. The amount of heat required to raise the temperature of any number n of grams of water from 15°C. to 16°C. is equal to n calories.

It is found that the amount of heat required to raise the temperature of one gram of water one degree at any temperature, for example from 25°C. to 26°C. or from 60°C. to 61°C., is nearly equal to the amount required to raise it from 15°C. to 16°C. If two equal masses of water, one at a temperature 16°C. and the other at a temperature 14°C., are mixed together, the mixture is found to have a temperature nearly equal to 15°C. We suppose, as it is natural to do, that when a hot body and a colder body are put in contact, the heat lost by the hot body to the cold body is equal in amount to the heat gained by the cold body from the hot body and that the amount of heat gained by a body when its temperature rises from $t_1°$ C. to $t_2°$ C. is equal to the amount of heat lost by the body when its temperature falls from $t_2°$ C. to $t_1°$ C. According to this the heat lost by the water at 16°C. is equal to the heat gained by the water at 14°C. Since the temperatures of the two equal masses of water are both changed by 1°C. it follows that the heat required to raise a gram of water from 14°C. to 15°C. is nearly equal to that required to raise it from 15°C. to 16°C. If equal masses of water at any temperatures are mixed together it is found that the temperature of the. mixture is nearly equal to the mean of the temperatures before mixing. Thus if 1000 grams of water at 50°C. are mixed with 1000 grams at 10°C. the mixture will be at nearly 30°C. This shows that the amount of heat required to raise the temperature of one gram of water one degree is nearly the same whatever the initial temperature of the water. Exact experiments show that this is not exactly true, but it is sufficiently nearly true for most purposes.

If a mass of m grams of water has its temperature changed from $t_1°$C. to $t_2°$C. the amount of heat required for the change is therefore approximately equal to $m(t_2 - t_1)$ calories. Quantities of heat are very often measured by finding how much they change the temperature of a known mass of water.

Suppose for example that a piece of lead weighing 1000 grams at a temperature of 100°C. is put into 1000 grams of water at 15°C. If the water is stirred up, the lead and water soon arrive at a temperature of about 17·5°C. It appears that the heat given out by 1000 grams of lead when it cools from 100° to 17·5°, or through a range of 82·5 degrees, is equal to the heat required to warm 1000 grams of water from 15° to 17·5° or only 2·5 degrees. It is clear that a mass of lead requires much less heat to warm it through any range of temperature than an equal mass of water. The amount of heat required to raise the temperature of unit mass of any substance one degree is called the specific heat of the substance. The product of the mass of a body and its specific heat is called its capacity for heat; it is also sometimes called the water equivalent of the body. If s denotes the specific heat of any substance, then the amount of heat required to raise the temperature of a mass m of it from t_1 to t_2 is equal to $sm\,(t_2 - t_1)$ calories. In the experiment just described we have therefore

$$1000\,(17\cdot5 - 15) = s\,1000\,(100 - 17\cdot5),$$

where s is the specific heat of the lead. This equation gives $s = 0\cdot03$ approximately. The following table gives the values of the specific heats of some substances at ordinary temperatures:

Substance			Specific Heat
Copper	...	...	0·095
Iron	...	...	0·114
Lead	...	...	0·031
Aluminium	...	...	0·219
Sodium chloride		...	0·313
Ice ...	...	...	0·502
Quartz	...	...	0·191
Water	...	...	1·000

Specific heats are often found experimentally by what is called the method of mixtures. This consists in mixing a mass of the substance at a known temperature with a quantity of water and finding the change in the temperature of the water. To obtain exact results a number of details must be attended to. The temperature of the vessel containing the water also changes, so that the heat required for this must be allowed for. During the experiment there may be some exchange of heat between the vessel of water and surrounding bodies, and the substance may lose heat while it is being put into the water. The vessel used to contain the water in such experiments is called a calorimeter. Fig. 13 shows a vertical section of a simple form of calorimeter. $AAAA$ is a double-walled cylindrical vessel open at the top. This is made of brass, nickel plated and polished inside and out. The space between the walls is filled with water, the temperature of which is measured with a thermometer T' passed through a cork as shown. The vessel $AAAA$ is called the water jacket. The opening at the top of it can be closed with a round wooden cover PP. The calorimeter proper is a thin-walled cylindrical vessel CC made of brass or copper, nickel plated and polished. It rests inside the water jacket on three wooden pegs or corks. It is about three-fourths filled with water, the temperature of which is measured with a thermometer T passed through a cork in a hole in the wooden cover. The water in the calorimeter can be stirred up with a stirrer S, consisting of a brass or copper rod attached to a circular ring. Bodies can be put into the calorimeter through a hole R in the wooden cover which can be closed with a cork. The calorimeter is designed to prevent as far as possible loss or gain of heat between it and surrounding bodies, and to enable any such loss or gain to be estimated so that allowance can be made

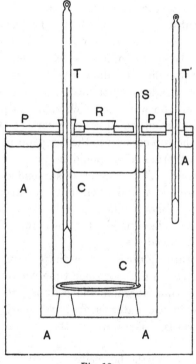

Fig. 13.

for it. It is found that a warm body loses heat to cooler surrounding bodies at a rate proportional to the difference between its temperature and the temperature of the surrounding bodies. The calorimeter is therefore enclosed in the water jacket, the temperature of which is observed and should remain almost constant. The calorimeter is supported by wooden or cork pegs because these conduct heat very badly. It is found that polished surfaces lose heat much less readily than dull surfaces, so that the outside of the calorimeter and the inside of the water jacket are brightly polished. The top is covered with a wooden lid to prevent air currents blowing over the calorimeter, which would tend to cool it down. The water in the calorimeter is always well stirred before observing its temperature; if this is not done, parts of the water may be much hotter than other parts.

To find the specific heat of a substance a known mass of it is heated to a definite temperature and then put into the calorimeter. If the substance is a solid body which is not acted on by water it may be used in the form of a hollow cylinder with a number of holes bored through its sides so as to expose a large surface to the water and to allow free circulation of the water through it. The cylinder may be heated in a double-walled vessel, or heater, between the walls of which steam is passed. The top and bottom of the vessel are closed with corks, and a thermometer passed through the top cork gives the temperature of the cylinder. The bulb of this thermometer should be placed in the middle of the hollow cylinder. The cylinder is suspended by a thread in the middle of the heater, and when its temperature has remained constant for some time the heater is brought above the calorimeter and the cylinder quickly let down into the water. The heater is then removed.

The temperature of the water in the calorimeter is observed every minute or half minute for some time before the hot body is put into it. The time at which the body is put in is noted and the temperature of the water is observed every half minute for some time afterwards. The water in the calorimeter is kept well stirred.

Suppose that before the hot body was put in, the temperature of the water fell at the rate of $\alpha°$C. per minute. Let the temperature of the water just before be $t_1°$C. Suppose that after the hot body was put in, the temperature of the water rose in a time T minutes to $t_2°$C. and then began to fall at the rate of $\beta°$C. per minute. Suppose also that during the whole experiment the water in the water jacket was at $t_3°$C. The rate at which the temperature of the water in the calorimeter falls is nearly proportional to the difference between it and the temperature of the water jacket.

During the rise from t_1 to t_2 the average temperature difference is the mean of $t_1 - t_3$ and $t_2 - t_3$. At t_1 the rate of fall is α, and at t_2 the rate is β. During the rise from t_1 to t_2 the average rate of fall is therefore roughly equal to $\frac{1}{2}(\alpha + \beta)$, so that the total fall during the time T is nearly equal to $\frac{1}{2}(\alpha + \beta) T$. If there had been no loss of heat from the calorimeter its temperature would therefore have risen to $t_2 + \frac{1}{2}(\alpha + \beta) T$. Let $t_2' = t_2 + \frac{1}{2}(\alpha + \beta) T$. t_2' is called the corrected final temperature of the water in the calorimeter. Let t be the temperature of the hot body before it was put into the calorimeter. The heat given out by the hot body is equal to $sm(t - t_2)$ where s is its specific heat and m its mass. The heat received by the water and calorimeter, etc. is equal to $(w + w')(t_2' - t_1)$ where w is the mass of the water in the calorimeter and w' denotes the total heat capacity of the calorimeter, stirrer and thermometer. We have therefore

$$sm(t - t_2) = (w + w')(t_2' - t_1).$$

By means of this equation s can be calculated. The heat capacity of the calorimeter, stirrer and thermometer can be calculated by adding up the products of their masses and specific heats, or it can be found experimentally. In finding it experimentally the calorimeter is about one-fifth filled with hot water and its temperature observed. Let it be t_1. Enough cold water at t_2 is then poured in until the calorimeter is about three-quarters full. Let the mixture take up a temperature t_3. Then we have

$$(w + w')(t_1 - t_3) = m(t_3 - t_2).$$

Here w denotes the mass of the water in the calorimeter at first and m the mass of water poured in. It is best to start with hot water, at nearly $100°$ C., in the calorimeter and to pour in cold water, because the total capacity for heat w' of the calorimeter, stirrer etc., is usually very small.

The specific heat of a liquid can be found in the same way as that of a solid. The liquid is enclosed in a metal bottle or a glass bulb. The heat given out by the bottle or bulb must of course be allowed for. Another way is to use the liquid, the specific heat of which is required, instead of water in the calorimeter, add to it a hot body of known heat capacity and observe the rise of temperature.

Dulong and Petit discovered that for many elements the product atomic weight × specific heat is nearly constant. This rule is called Dulong and Petit's law. The following table gives some values of the product in question. It is not exactly constant.

Element		Atomic Weight	Specific Heat	Product
Sodium	...	23	0·293	6·75
Sulphur	...	32	0·178	5·68
Iron ...	...	56	0·1140	6·37
Bromine	...	80	0·0843	6·74
Silver	...	108	0·0570	6·16
Lead	...	207	0·0314	6·50

It is found that the specific heats of most substances are smaller at low temperatures than at high temperatures. At very low temperatures near − 273° C., which is the lowest possible temperature, the specific heats become very small. As the temperature rises the specific heats rise at first more or less rapidly and then more slowly, and at higher temperatures become nearly constant. Dulong and Petit's law applies to the nearly constant values attained at the higher temperatures. The specific heats of metals have nearly reached their constant values at ordinary temperatures, but those of carbon, silicon and some other bodies do not become nearly constant until very high temperatures are reached. For example the specific heat of graphite, a form of carbon, has the following values:

Specific Heat	Temperature
0·152	0° C.
0·443	600
0·453	800
0·467	1000

The specific heat of platinum between 0° C. and 1200° C. is given by the equation

$$s = 0.0317 + 0.000012t,$$

where t denotes the temperature.

REFERENCES

Heat, Poynting and J. J. Thomson.
Advanced Practical Physics, Watson.

CHAPTER VI

CHANGE OF STATE

Solid—Liquid

IN previous chapters we have considered some of the chief properties of solids, liquids, and gases. We shall now discuss changes from one of these states into another. If a block of ice at a temperature below 0°C. is brought into a warm room at, say, 20°C., the temperature of the ice begins to rise. This rise may be observed if a thermometer is put into a hole bored into the block. The temperature of the block rises till it reaches 0°C., and then remains constant; but the ice begins to melt into water. The block of ice is receiving heat all the time from the hotter bodies surrounding it, but in spite of this its temperature remains 0°C.; the heat is used up in melting the ice. The heat supplied to the ice is said to be rendered latent. It is found that to melt one gram of ice at 0°C. into water also at 0°C. requires 80 calories to be supplied. If a mixture of crushed ice and water is put in a cold place at a temperature below 0°C., the mixture loses heat to the surrounding colder bodies, but it does not get any colder. The water gradually freezes and the temperature of the mixture remains constant at 0°C. When all the water has frozen, the ice begins to cool down. When water freezes each gram gives out 80 calories. Thus we see that the heat rendered latent when ice is melted is given out again when the water formed freezes back into ice. It is found that when other solid bodies are melted heat is rendered latent in the same way as with ice. The latent heat of any substance is the amount of heat required to convert unit mass in the solid state at its melting point into liquid at the same temperature. An intimate mixture of ice and water is always at the melting point of ice. If heat is supplied to

the mixture some of the ice melts, and if heat is taken away some
of the water freezes. If no heat enters or leaves the mixture the
amounts of ice and water in it remain constant. We say that ice
and water are in equilibrium with each other at the temperature
0°C. The melting point of a substance may be defined as the
temperature at which the solid and liquid states can exist together
in equilibrium. To raise the temperature of a mixture of the
solid and liquid states of a substance above the melting point it is
necessary first to melt the solid; and to cool the mixture below the
melting point it is necessary first to freeze all the liquid. The
solid and liquid cannot exist together in equilibrium except at
the melting point.

To find the melting point of any substance the best way is to
prepare an intimate mixture of the solid and liquid states and
find its temperature with a thermometer. For example, to find
the melting point of tin heat a quantity in a nickel crucible over
a Bunsen burner until it has all melted. Then turn down the
flame so that the liquid cools slowly, and stir it up with a
thermometer. After a time the liquid will begin to solidify; and,
by stirring, an intimate mixture of the solid and liquid can be
obtained. While such a mixture exists the thermometer will
indicate a constant temperature, which is the melting point of
the tin.

The latent heat of ice can be found experimentally by putting
a known mass of dry ice at 0°C. into a calorimeter containing
warm water. A good way is to wrap up a piece of clear ice in
blotting paper and weigh the ice and paper together. Then slip
the ice out of the paper into the calorimeter and weigh the paper
again. The paper absorbs any water formed from the ice, so that
the difference between the two weights gives the weight of the
ice put into the calorimeter. The water in the calorimeter is
stirred up until the ice has all melted. Let t_1 be the temperature
of the water before the ice was put in and t_2 the temperature after
the ice had melted. Then we have

$$mF + mt_2 = (w + w')(t_1 - t_2),$$

where m is the mass of the ice and F its latent heat, w the mass
of water in the calorimeter, and w' the heat capacity of the

calorimeter, stirrer, and thermometer. The heat lost by the water and calorimeter, etc. is $(w + w')(t_1 - t_2)$; the heat required to melt the ice into water at $0°$ C. is mF; and the heat required to warm the water formed from the ice from $0°$ C. to the final temperature is mt_2.

The latent heat of other substances can be found in a similar way by putting a piece of the solid into some of the liquid and observing the resulting change of temperature when the solid has all melted. If s denotes the specific heat of the liquid and s' that of the solid, then we have

$$mF + sm(t_2 - t_0) + s'm(t_0 - t_s) = (sw + w')(t_1 - t_2).$$

Here m is the mass of the solid, F its latent heat, t_2 the final temperature of the liquid, t_0 the melting point of the solid, t_s the initial temperature of the solid, w the mass of the liquid, w' the heat capacity of the calorimeter, etc., and t_1 the initial temperature of the liquid. This method is suitable only for substances which melt at temperatures below about $40°$C. In all such experiments it is necessary to correct the observed temperatures so as to allow for the heat gained or lost by the calorimeter from the water jacket. This is done in the way described in the previous chapter.

The following table gives the latent heats and melting points of some solids :

Substance	Latent heat of fusion in Calories per gram	Melting point
Ice	80	$0°$ C.
Phosphorus ...	5	44
Sulphur ...	9·4	115
Bromine ...	16·2	− 7·3
Sodium thiosulphate	37·6	9·9
Tin	14	230
Silver	21	1000
Lead	5·4	325
Benzene ...	30	5·4
Phenol	25	25·4

The change from the liquid state into the solid state is usually accompanied by a change of volume. For example, unit volume of water becomes about 1·09 unit volumes of ice. Thus the density of the solid at the melting point is usually different from the density of the liquid at the same temperature. The following table gives some values of the densities in grams per c.c. at the melting points :

Substance	M. pt.	Density of solid	Density of liquid
Water	0° C.	0·91603	0·999868
Lead	325	11·01	10·65
Sodium	97·6	0·952	0·929
Mercury	− 38·9	14·19	13·69

In most cases the density of the liquid is less than that of the solid, so that the solid sinks in the liquid. Water is an exception to this rule, so that ice floats on water. It is found that the melting points of substances, such as ice, the volume of which changes when they melt, depend on the pressure at which the melting takes place. Usually the pressure is the ordinary atmospheric pressure, about 76 cms. of mercury. At this pressure ice melts at 0°C. At a pressure of 100 atmospheres or about 1500 pounds weight per square inch ice melts at − 0·74°C. instead of at 0°C. The melting points of substances which expand when they melt are made higher by great pressure, while those which contract like ice have their melting points lowered.

The change of volume when ice melts can be measured with the apparatus shown in Fig. 14. This consists of a glass bulb, of about 100 c.c. capacity, blown on a tube about 0·5 cm. in diameter. About 30 c.c. of water are put in the bulb, and it is then filled up, to a point in the tube, with paraffin oil. The bulb is put in a mixture of crushed ice and hydrochloric acid which acts as a freezing mixture and gives a temperature of about − 20°C. The water then freezes and its expansion causes the oil to rise considerably in the tube. If the rise and the diameter of the tube are measured the change of volume can be easily calculated. Water pipes are often burst in cold weather when the water in them freezes, on account of the greater volume of the ice formed. If a thick walled cast iron bottle, closed by an iron plug screwed in, is filled with water and put in a freezing mixture, the bottle will burst when the water freezes. If the pressure required to burst the bottle is, say, 200 atmospheres, then the bottle and the water in it have to be cooled down to − 1·48° C. before the freezing can go on sufficiently to produce this pressure.

Many substances such as glass, wrought iron and sealing

wax have not a definite melting point. When they are heated
they gradually get softer as the temperature rises, and finally
become liquid by a continuous process. When gradually cooled

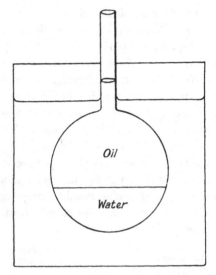

Fig. 14.

the liquid becomes more and more viscous until it is brittle and
behaves like a solid. It is difficult to distinguish between an
extremely viscous liquid and a soft solid.

At the melting point of a substance the solid and liquid states
can exist together in equilibrium. If the temperature at the surface of separation between
solid and liquid is not the melting point, then either the solid
melts or the liquid freezes at the surface according as the temperature is above or below the melting point. If the temperature is
above the melting point the solid melts and absorbs heat which
tends to keep down the temperature, so that it is not possible to
raise the temperature appreciably above the melting point. When
the solid is melting rapidly, however, the temperature at the surface
of separation must be slightly above the melting point, for if it
was exactly at the melting point the solid and liquid would be
in equilibrium and the solid would not melt. It is found that

Velocity of
Solidification.

liquids can be cooled down to temperatures considerably below
the melting point without solidifying, provided no solid is present.
This phenomenon is called super-cooling. If a glass tube a few
millimetres in diameter and say 15 cms. long, closed at one end, is
filled about two-thirds full of liquid carbolic acid, phosphorus,
sulphur, salol or other substance, it can be cooled below the melting
point without the liquid solidifying. The melting point of salol is
41°C., but liquid salol in a tube like that just described can be
cooled down to 10°C. without solidifying. If a small piece of
solid salol is held in the super-cooled liquid close to its surface,
then solidification takes place at the surface of the solid so that
it grows down the tube with a definite velocity which is called the
velocity of solidification. This velocity depends on the temperature of the surface of the solid. As the super-cooling is increased
the velocity increases at first rapidly, but then more slowly, and
then remains constant over a considerable range of temperature.
The following table gives the velocities of solidification of salol at
several temperatures :

Temperature	Velocity, mms. per minute
41·0°C.	0·0
36·6	0·4
34·1	1·8
33·2	2·3
28·1	3·8
24·4	4·0
20·3	4·1
15·1	4·1

Other substances behave in a similar way. For example, benzoic
anhydride, which melts at 42°C., gives a maximum velocity of
solidification of 35 mms. per minute. Water can be super-cooled
considerably ; its velocity of solidification is large, but has not yet
been determined accurately.

CHAPTER VII

CHANGE OF STATE

Liquid—Vapour

IF a dish of water is left standing in the open air the water gradually disappears. It is converted into water vapour which mixes with the air and is carried away by the wind. To study this process more closely the apparatus shown in Fig. 15 may be used. B is a large three-necked bottle. At A a tube with a cock in it leads into the bottle. This can be connected to an air pump and the bottle exhausted. At C is a tap funnel containing some liquid such as water, alcohol or ether. At D a closed mercury manometer MN is connected to the bottle and indicates the pressure of the gas in the bottle. Suppose that the bottle is almost completely exhausted so that the mercury stands at the same level in the tubes M and N. The pressure in N is supposed to be zero. If now one drop of the liquid, ether say, is let into the bottle by opening the tap of the tap funnel for a moment,

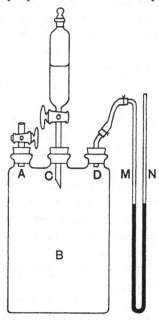

Fig. 15.

it is found that the pressure rises by a small amount and the drop of ether all evaporates and so becomes invisible. The bottle is then full of ether vapour at a small pressure which is indicated by the manometer. If now another drop is let in, it also evaporates and

the pressure is about doubled. If a third drop is let in, it evaporates and the pressure becomes about three times that due to the
first drop. If more drops of ether are let in, the pressure goes on
increasing and the drops evaporate until a certain definite pressure
is reached; and then on letting in more ether no further increase
in the pressure takes place, and the ether does not evaporate but
remains in the liquid state on the bottom of the bottle. The final
value of the pressure depends on the temperature.

If the temperature of the bottle is 20° C., the final pressure is
about 44 cms. of mercury when the liquid used is ether. If alcohol
is used the final value is 4·4 cms. of mercury, and if water is used
it is only 1·74 cms. of mercury at 20° C. It appears that a liquid in
a closed vessel evaporates until the vapour attains a certain pressure,
and then the liquid and vapour remain together in equilibrium
and no further evaporation takes place. The pressure of the
vapour at which the liquid and vapour can exist together in
equilibrium is called the vapour pressure of the liquid. The
vapour pressure of a liquid depends on the temperature of the
liquid. It increases rapidly as the temperature rises. Instead
of starting with the bottle B completely exhausted we may start
with some gas such as air or hydrogen in it. Suppose at the
start the bottle contains air at 20 cms. pressure. If now ether is
let in it is found that the pressure rises just as it did before, but
the final pressure attained is nearly 64 cms. of mercury instead of
44 cms. The increase in the pressure due to the ether is nearly the
same as before. It appears therefore that in the presence of a gas
a liquid evaporates until the pressure due to its vapour is nearly
the same as it would have been if the gas had not been there.
Exact experiments show that the presence of the gas very slightly
diminishes the vapour pressure, but the effect is negligible for
most purposes.

Suppose we start with the bottle B completely exhausted and
let in ether until the maximum pressure is reached and no more
ether evaporates. Then pour mercury into the tap funnel and
allow it to run into the bottle. Mercury gives off practically
no vapour. In this way the volume of the space occupied by the
vapour in the bottle can be diminished. It is found that as the
mercury is run in the pressure indicated by the manometer remains

unchanged. We can fill the bottle nearly full of mercury without changing the pressure, but the amount of liquid ether in the bottle increases as the volume of the ether vapour is diminished. Thus when the volume of a vessel containing the vapour of a liquid at its maximum pressure is diminished, the pressure does not increase but the vapour condenses into liquid.

If we begin to run in mercury when the amount of ether let into the bottle is not enough to give the maximum pressure, then the pressure increases as the mercury runs in until it reaches the maximum pressure. If more mercury is run in, the pressure then stays constant and the ether vapour condenses. It appears therefore that at a given temperature only a certain amount of ether in the form of vapour can exist in a given volume. If the volume is diminished the vapour begins to condense into liquid when the amount per c.c. has attained the maximum value. The space is then said to be saturated with ether and the vapour is said to be a saturated vapour. A saturated vapour is a vapour which can exist in equilibrium with the liquid. The vapour-pressure of a liquid may be said to be the pressure of its saturated vapour. In the absence of any liquid a vapour may have a pressure less than the saturated pressure; it is then called an unsaturated vapour.

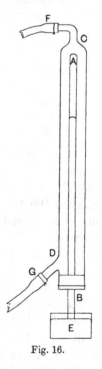

The vapour pressure of a liquid can be found with the apparatus shown in Fig. 16. AB is a glass tube closed at A. It is filled with mercury and inverted in a vessel of mercury E so as to form a barometer. This tube AB is enclosed in a wider tube CD through which hot water or steam can be passed by means of the tubes F and G. If the tube AB is properly filled with mercury there will be a good vacuum at the top of it and the mercury in it will

Fig. 16.

stand above that in E at a height equal to the height of the barometer. To find the vapour pressure of a liquid such as water a few drops of it are put into the bottom of AB beneath the

mercury by means of a small pipette, like those used for filling
fountain pens. The liquid floats to the top of the mercury column
and the vacuum is filled with the vapour. Enough liquid must be
introduced for some of it to remain in the liquid state. The
vapour at the top of AB is then saturated and its pressure forces
down the mercury a distance h such that the vapour pressure is
equal to the pressure due to a column of mercury of height h.
By passing water at known temperatures through CD the de-
pression of the mercury can be found at different temperatures.
The following table gives the vapour pressure of some liquids at
different temperatures:

Vapour Pressure in mms. of Mercury.

Temperature	Water	Alcohol	Ether	Mercury
0° C.	4·58	12·5	184·7	—
10	9·18	24·0	292	—
20	17·41	44·2	442	—
50	92·17	219·9	1276	0·015
70	233·79	541	2294	0·052
100	760·00	1695	4859	0·270
150	3581·0	7321	13281	2·68
200	11688·0	22164	—	17·02
300	67620·0	—	—	246·70

If an open vessel containing water is heated over a flame the
temperature of the water rises steadily until it reaches about
100° C. The water then boils and its temperature remains constant at 100° C. Bubbles of vapour are formed beneath the surface
of the water and these grow larger and rise to the surface where
they burst. The more rapidly heat is supplied to the water the
more vigorously it boils. When boiling, the water is being converted into steam or water vapour. In order that a liquid may boil
it must be heated till its vapour pressure is equal to the pressure
of the gas above it. At 76 cms. of mercury pressure water boils
at 100° C. If the atmospheric pressure is less than 76 cms. water
boils below 100° C. and if the pressure is greater than 76 cms. it
boils above 100° C. A very good way to find the vapour pressure
of a liquid at different temperatures is to find its boiling point at
different pressures. This can be done with the apparatus shown
in Fig. 17. The liquid is contained in a metal boiler B heated by

a burner P. The temperature of the vapour above the boiling liquid is measured with the thermometer T. The boiler is connected to a tube GH which slopes upwards. This tube passes through a wider tube CC through which cold water is passed by means of side tubes at E and F. The vapour from the boiler condenses in GH and the liquid formed runs back into the boiler. At H the tube GH leads into a large vessel V. This is connected to a mercury manometer M and to an air pump through the tube A. By means of the pump the air pressure in the apparatus can be brought to any desired value, which can be measured with the

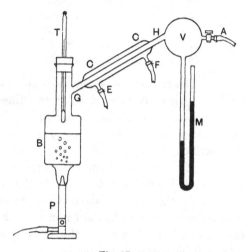

Fig. 17.

manometer. The liquid in the boiler is kept boiling steadily and the temperature of its vapour is read off on the thermometer. For example, if the liquid used is water and the pressure is adjusted to 358·1 cms. of mercury it is found that the water boils at 150° C. The vapour pressure of water at 150° C. is therefore 358·1 cms. of mercury.

It is found that if the water is not pure but has salts or other substances dissolved in it the temperature of the boiling solution is slightly higher than that of pure water boiling at the same pressure. The temperature of the vapour rising from the boiling solution however is found to be the same as that of

the vapour from pure water. It is therefore best to measure
the temperature of the vapour above the boiling liquid when
finding the boiling point of a liquid at any pressure.

The following table gives the boiling points of water at several
pressures near to 760 mms. of mercury:

Pressure, mms. of mercury	Boiling point
740	99·255° C.
745	99·443
750	99·630
755	99·815
760	100·000
765	100·184
770	100·366
775	100·548
780	100·728

When it is necessary to test a thermometer to see if the
upper fixed point is correctly marked on it the thermometer is
immersed in the steam from boiling water in a suitable vessel
from which the steam can escape freely. The temperature indicated by the thermometer is then observed and also the height
of the barometer. If the height of the barometer is 76 cms. then
the thermometer should indicate 100°C., but if the barometer is
not at 76 cms. then the thermometer should show the boiling point
of water at the pressure indicated by the barometer. For example,
if the barometer stands at 745 mms. the thermometer should
indicate 99·443 if it is correctly graduated.

When an open vessel of water or other liquid is heated over
Latent Heat of   a flame so that it is kept boiling, its tempera-
Evaporation.   ture does not rise although heat is continually
entering it. This heat is used up in converting the liquid into
vapour. To condense the vapour back into liquid form it is
necessary to take out of it an amount of heat equal to that which
was used in converting the liquid into vapour. The amount of
heat required to convert a unit mass of a liquid at any temperature
into vapour at the same temperature is called the latent heat of
evaporation of the liquid at that temperature. To convert one
gram of water at 100° C. into steam at 100° C. requires 536 calories.

The latent heat of evaporation of a liquid such as water or

alcohol may be found with the apparatus shown in Fig. 18.
The liquid is boiled in a boiler B and its vapour passes through a
rubber tube A into a trap C. This trap serves to catch any liquid
which may come over from B along with the vapour. The dry
vapour from the trap enters a spiral tube D which leads to a box
E. An open tube F leads from this box to the air. The spiral

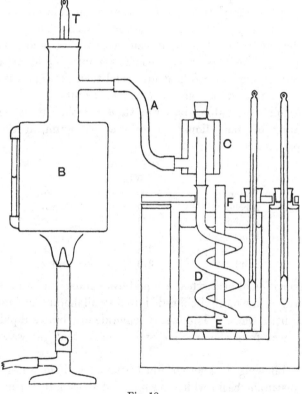

Fig. 18.

and box are immersed in cold water in a calorimeter as shown.
The trap is first removed from above the spiral condenser and
the boiler heated till the liquid boils slowly and steadily. The
trap is then emptied and put on the condenser. After a short
time when the calorimeter temperature has risen about 10° or
15°C. the trap is removed. Let the initial temperature of the

water in the calorimeter be t_1, the final temperature be t_2 and the corrected final temperature t_2'. The heat received by the calorimeter, condenser, stirrer, and thermometer is equal to that given out by the vapour in condensing. We have therefore

$$mL + ms\,(t - t_2) = (w + w')\,(t_2' - t_1).$$

Here m is the mass of the vapour condensed, L the latent heat of the vapour, s the specific heat of the liquid formed, t the temperature of the vapour, w the mass of water in the calorimeter and w' the heat capacity of the calorimeter, condenser, stirrer, etc. m is found by weighing the spiral condenser before and after the experiment. The vapour first condenses into liquid at a temperature t, giving out mL heat units and then the liquid is cooled down from t to t_2 giving out $ms\,(t - t_2)$ heat units.

The following table gives the values of the latent heats of evaporation of some liquids in calories per gram, at the temperatures stated:

Liquid	Temperature	Latent Heat
Water	0° C.	598
Water	100	536
Alcohol	0	237
Alcohol	78	204
Ether	35	90
Mercury	358	68
Sulphur	316	362

The absorption of heat when a liquid evaporates can be illustrated by pouring some methyl chloride into a small metal crucible standing in a little water. The methyl chloride evaporates rapidly and absorbs heat from the crucible and water so that the water soon freezes.

The following is another illustration. A little water is put into a platinum basin which is supported above a dish containing strong sulphuric acid. The basin and dish are covered with a bell jar which is then exhausted by means of an air pump to a pressure less than one millimetre of mercury. The water then boils and its vapour is absorbed by the acid. The rapid evaporation cools the water so that it soon freezes although it is boiling.

The volume of vapour formed when a liquid evaporates is usually much greater than that of the liquid. The density of the vapour is therefore much smaller than that of the liquid. For example one gram of water gives 1663 cubic centimetres of saturated vapour at 100°C.

The relation between the pressure and volume of a quantity of any substance can be represented on a diagram by taking distances measured vertically upwards from a horizontal line to represent the pressures and the distances measured horizontally from a vertical line to represent the corresponding volumes. In Fig. 19 the point P represents the state of a substance when its volume

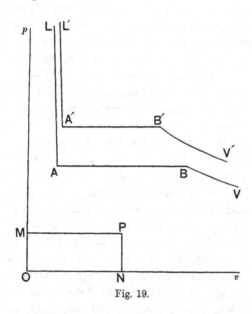

Fig. 19.

is represented by MP or ON and its pressure by NP or OM. Suppose we represent in this way the relation between the pressure in a vessel containing a liquid and its vapour and the volume of the vessel. If the volume is diminished some of the vapour condenses, and if the temperature is kept constant the pressure remains constant. So long as both liquid and vapour are present the relation between pressure and volume at constant temperature is therefore a straight horizontal line like AB. If the volume is diminished until all the vapour has condensed and nothing but the liquid remains in the vessel, then increasing the pressure will produce only a small diminution of volume because liquids are very nearly incompressible. If A represents the state when all the vapour has condensed, then the line showing the relation between pressure and volume at constant temperature for volumes less than that corresponding to A will be nearly vertical, like AL. If the volume is increased until all the liquid has evaporated at the constant temperature, then on further increasing the volume the pressure will fall because the vapour will then not be saturated. Suppose that B represents the state when all the liquid has just evaporated and there is nothing but saturated vapour present in the vessel. For volumes greater than that corresponding to B a curve like BV will represent the relation between p and v at constant temperature. The complete curve $LABV$ is called an isothermal curve. The part LA represents the state of the substance when it is all liquid, the part AB when it is partly liquid and partly vapour and the part BV when it is all vapour.

It is found that the relation between the pressure and the volume is represented by a curve like $LABV$ at all temperatures below a certain temperature which is different for different substances. As the temperature is raised the pressure of the saturated vapour rises and its density increases and the volume of the liquid increases. Consequently at higher temperatures the horizontal part of the curve is higher up and shorter, like $L'A'B'V'$. The part like AB gets shorter as the temperature rises until it disappears and then there is no longer a stage in which the substance separates into the liquid and gaseous states. The highest temperature at which this separation can take place is called the *critical temperature* of the substance. Above this temperature the substance cannot be liquefied, that is, it cannot be made to separate into two different states with a surface of separation between them.

The existence of the critical temperature can be demonstrated with the apparatus shown in Fig. 20. This consists of a strong glass tube AB closed at A and joined to a bulb at B. The bulb and tube are filled with pure carbon dioxide gas and the bulb is immersed in mercury contained in a steel tube CD. At the lower end of CD there is a screw by means of which the mercury can

be forced into the bulb and the tube AB so that all the gas is
compressed into the upper part of AB.
The tube AB is surrounded by a wider
tube containing water, the temperature
of which can be measured with a thermometer. If the temperature is below 31°C.,
then on compressing the gas in AB sufficiently it separates into two parts, with
a distinct surface of separation between
them.   The lower part is liquid and the
upper part gas or vapour.   By screwing
up the mercury the substance can be
completely converted into liquid and the
length of AB which it then occupies observed.   By slowly lowering the mercury
the substance can be just all converted
into gas and the length this occupies
observed. If this is done at a series of
temperatures it is found that, as the
temperature gets near to 31° C., the
volume of the liquid gets greater and
that of the vapour less until very close to
31°C. they become nearly equal, and at or
above 31°C. the substance does not separate into two states at all.   If the volume
is adjusted so that the space above the
mercury is about two-thirds filled with
liquid, then on slowly raising the temperature it can be seen that the surface of
separation between the liquid and vapour
gets gradually fainter and harder to see
until at 31°C. it disappears. The pressure,
density and temperature at which the
density of the liquid becomes equal to the
density of the vapour are called the critical
values of these quantities. The apparatus

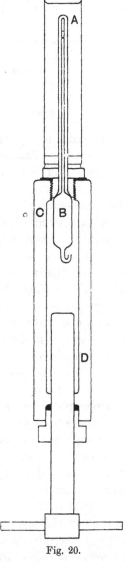

Fig. 20.

described enables the critical temperature to be found and by
connecting it to a manometer and graduating the tube into

known volumes, the critical density and pressure can also be
determined. The following table gives the values of the critical
quantities for several substances:

Substance	Temperature	Pressure in Atmospheres	Density (Grams per c.c.)
Ether	195° C.	36	0·3
Carbon dioxide	31	77	0·45
Water	360	200	0·5
Oxygen	−119	56	0·65
Hydrogen	−238	15	0·05

In order to liquefy a gas it must be cooled below its critical
temperature. Thus oxygen cannot be liquefied above − 119° C.
and carbon dioxide above 31° C. All gases, including hydrogen and
helium have been liquefied. The gases like air, oxygen, hydrogen and helium which are difficult to liquefy because their critical
temperatures are low can be liquefied by an ingenious process
invented independently by Hampson and Linde. This process
depends on the fact that when a gas expands without doing work it
is cooled slightly. All gases show this effect at low temperatures.
This small cooling effect is believed to be due to the molecules of
the gas attracting each other so that when the volume of the gas
increases and the molecules separate to greater distances apart
their velocity is diminished. In Hampson's apparatus for producing liquid air, the air is compressed by a pump to about 180
atmospheres pressure. Compressing air makes it hot because work
is done on it and so it has to be cooled by passing it through spirals
of copper pipe immersed in cold water. The air at 180 atmospheres
pressure is freed from carbon dioxide and water vapour by passing
it through a vessel containing sticks of solid caustic potash; it
then enters a long coil of narrow copper tubing contained in a
box surrounded by layers of felt. The air is allowed to escape
from the copper tubing through a valve and its pressure falls to
slightly more than one atmosphere. The valve is near the bottom
of the box with the coil of tubing above it. The air, after escaping
through the valve, passes from the bottom to the top of the box in
between the coils of the copper tubing, and then goes from the top
of the box back to the pump where it is again compressed. The
expansion from 180 atmospheres pressure down to about one

atmosphere cools the air, and the cooled air as it passes up between the coils of tubing cools these down and so cools down the air moving through the tubing towards the valve. In this way the temperature at the valve is made to fall steadily until it reaches about $-190°$ C. when about five per cent. of the escaping air liquefies and the liquid collects at the bottom of the box. Liquid air is a colourless liquid like water. Its boiling point at 76 cms. of mercury pressure depends on the percentage of oxygen in it and varies between about $-180°$ and $-190°$ C. Hydrogen and helium can be liquefied by a similar process, but in the case of these gases it is necessary to cool the compressed gas to a very low temperature before letting it into the liquefying apparatus. Liquid hydrogen is a colourless liquid which boils at about $-253°$ C. at 76 cms. of mercury pressure. By allowing liquid hydrogen to evaporate at a pressure of only a few mms. of mercury it can be frozen into a transparent colourless solid which melts at about $-258°$ C. Liquid helium has a lower boiling point than liquid hydrogen.

REFERENCE

The Experimental Study of Gases, Travers.

CHAPTER VIII

CONVECTION AND CONDUCTION

In this chapter we shall discuss the two ways in which heat can move from one point to another. If a hot body is moved from one place to another its heat goes with it. This way of moving heat is called convection. If a vessel of water is heated from below, the hot water at the bottom is less dense than the colder water higher up. The hot water therefore floats up towards the top and the cold water sinks. A circulation of the water is thus set up, so that all parts of the water in turn come near to the bottom and get heated there. This circulation causes the water in the vessel to get hot much more quickly than it would if the water remained at rest. If a glass beaker full of water is heated from below with a small flame, the circulation of the water can be easily seen if a few crystals of potassium permanganate are put at the bottom of the beaker. The permanganate colours the water as it passes by the crystals and the stream of coloured water rising up the middle of the beaker and falling down its sides can be clearly seen. A similar circulation is produced when a closed vessel full of any gas is heated from below.

Convection.

Buildings are often warmed in cold weather by means of hot water which is passed through coils of pipe in each room. In this way heat is transferred from the furnace where the water is heated to all parts of the building. Hot air is sometimes used instead of hot water. The cold air from outside passes through pipes, heated by a furnace in the basement, which lead from the furnace to each room. The air enters the room and then escapes by the doors and windows or through special openings to the outside. Thus there is a continuous circulation of the air. This circulation is kept up just like the circulation in a closed vessel heated below. The hot light air rises and the cold air outside falls. The draught up a chimney is produced in the same way. Suppose the height of a chimney is h and let the average density of the hot air in it be ρ. The pressure at the bottom of the chimney if the gases were at rest would be greater than that at the top by $h\rho g$. Let the average density of the cold air outside the chimney be ρ'. The pressure at the bottom of the chimney is greater than that at the top by $h\rho'g$. But ρ' is greater than ρ so that there is an unbalanced difference of pressure equal to $hg\,(\rho'-\rho)$ which drives the gases up the chimney. They move up with a velocity such that the resistance to the motion inside the chimney produces a pressure equal and opposite to $hg\,(\rho'-\rho)$. The resistance to the motion is due partly to the inertia of the gases and partly to their viscosity.

Conduction.

If one end of a metal bar is heated it is found that heat travels along the bar towards the other end so that the temperature of the whole bar rises. If one end of a bar is kept hot and the other surrounded by ice, then the ice will be melted at a certain rate, showing that there is a flow of heat through the bar from the hot end to the cold end. Such a flow of heat through a substance is called conduction of heat.

Some substances conduct heat much more readily than others. Metals, especially silver and copper, conduct the best, while substances like wood and woollen cloth conduct very much less then metals.

The flow of heat along a copper bar can be studied with the apparatus shown in Fig. 21. AB is a copper bar about one foot long and $2\frac{1}{2}$ inches in diameter. The end A is enclosed in a copper box through which steam or hot water is passed by means of two tubes as shown. In this way the end A is kept hot. The end B is also enclosed in a copper box through which a steady stream of cold water is passed, thus keeping B cool. Heat flows along the bar from A to B and warms the water flowing round the end B. The temperatures of the water before and after going through the bar at B are determined with two thermometers T_3 and T_4 which are passed through corks in small vessels through which the water flows as shown. The temperatures of the bar are observed at two points C and D by means of thermometers T_1 and

T_2 inserted into small holes bored into the bar. The whole apparatus is thickly covered with felt which is a very bad conductor of heat and therefore practically stops any heat from escaping. If the mass of cold water flowing per second, through the box at B, is m grams, then the heat it received per second is equal to $m (t_4 - t_3)$, where t_3 is the temperature of the water before and t_4 that after going through the box. If the temperature of the stream of water, flowing through the box at A, is varied, then it is found that the amount of heat, flowing along the bar, is

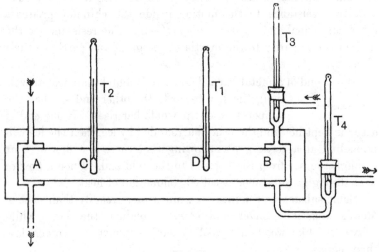

Fig. 21.

proportional to the difference between the temperature t_2 at C and the temperature t_1 at D. Thus it appears that the flow of heat is proportional to the change of temperature in any distance along the direction of the flow.

Suppose we have a thick slab of any substance bounded by two parallel planes. Let the temperature of one side of it be maintained constant at t_2 and the other at t_1. Let the thickness of the slab be d. The amount of heat which flows through an area A of the slab in a time T is given by the equation

$$H = kTA (t_2 - t_1)/d,$$

where k is a constant depending only on the nature of the material of which the slab is made. k is called the conductivity for heat of the material of the slab. If $A = 1$ sq. cm., $t_2 - t_1 = 1°$ C., $d = 1$ cm., and $T = 1$ second, we get $H = k$.

Thus the conductivity for heat of a substance may be defined as the amount of heat which flows in unit time through unit area of a slab of unit thickness when the difference between the temperatures of the two sides of the slab is one degree.

The apparatus shown in Fig. 21 above can be used to find the heat conductivity of the copper bar. We have

$$m\,(t_4 - t_3) = kA\,(t_2 - t_1)/d,$$

where A is now the area of cross section of the bar and d the distance from C to D. The following table gives some values of the conductivities for heat of different substances:

Substances	Conductivity
Copper	0·90
Iron	0·16
Lead	0·08
Aluminium	0·48
Silver	1·00
Glass	0·002
Sulphur	0·0006
Marble	0·005
Flannel	0·00004
Water	0·0014
Ether	0·0003
Air	0·000048
Hydrogen	0·00032

CHAPTER IX

HEAT A FORM OF ENERGY

IF a block of lead is hammered on an anvil with a heavy hammer it gets hot although the anvil and hammer were both cold. If a brass tube is mounted on a vertical shaft so that it can be rapidly rotated by means of a pulley and belt driven by a motor, and is squeezed between two wood blocks while it is rotating, it rapidly gets hot. If some water is put in the tube it soon begins to boil. Whenever two bodies are made to slide over each other so that work is done against the frictional resistance to the motion, heat is produced and the bodies get hot. For example, if the bearings of the axles on a railway car are not properly oiled they may become red hot when the car is running. In such cases work is done and heat appears. Thus the quantity of heat in a system is increased, although no heat has come in from outside, when some of the energy in the system is used up in doing work against frictional forces. For example, if a heavy body is allowed to fall on to a block of lead the lead gets hotter. The system consisting of the earth, the body and the block of lead has lost some potential energy but has gained some heat. It is found that when energy is used up in this way so as to produce heat the amount of heat produced is proportional to the amount of energy used up or to the work done. If an amount of work W is done and nothing but heat produced then

$$W = JH,$$

where H is the amount of heat produced and J is a constant. J is called the mechanical equivalent of heat, it is equal to the amount of work or energy which must be expended to produce one unit of heat. One calorie can be produced by the expenditure of 4.2×10^7 ergs. Hence W ergs $= 4.2 \times 10^7 \, H$ calories.

We conclude that heat is a form of energy. Thus when work is done and heat produced the total amount of energy remains unchanged; for the heat formed is an amount of energy equal to the energy used up in producing it. It is found that heat can be converted into other forms of energy. For example, if a steam engine is used to lift heavy weights it does work and increases the potential energy of the weights. It is found that the engine uses up a quantity of heat, so that more heat enters the engine than leaves it. This heat which disappears in the engine is converted into the potential energy of the weights which the engine lifts. If the total work done by the engine is W then

$$W = JH,$$

where H denotes the amount of heat which disappears in the engine and $J = 4 \cdot 2 \times 10^7$ ergs per calorie as before.

At one time it was supposed that heat was a material substance the total amount of which remained constant. When it was found that heat could be produced by the expenditure of work, this idea was given up. The relation between heat and work was first accurately investigated by Joule about the year 1840. The principle of the conservation of energy was first established on a satisfactory basis by Joule; for he showed that when work is done or energy used up in producing heat it has not disappeared, but has merely been converted into another form of energy. Many scientists before Joule believed that the total amount of energy remains constant and some made use of this idea, but Joule first carried out exact experiments which made its truth probable. Subsequent investigations have not brought to light any exception to this law. Energy may exist in many different forms such as heat, electrical energy, magnetic energy, gravitational energy, energy of elastic strains, chemical energy—but the total amount of it remains constant. It can be converted from one form to another but cannot be created or destroyed.

The amount of work required to produce a definite amount of heat can be measured approximately with the apparatus shown in Fig. 22. This apparatus is similar in principle to one form of apparatus used by Joule. A small copper calorimeter is supported on a block of ivory or other hard bad conductor of heat.

The block is mounted on a vertical shaft which can be made to rotate rapidly by means of a pulley and belt. The shaft carries a worm wheel which drives another wheel having 100 teeth. This serves to indicate the number of revolutions made by the shaft. The copper calorimeter consists of two conical vessels, one of which fits into the other. The inside one carries a wooden wheel to which a thread is fastened. The thread is wrapped round the wheel about once and then passes over a pulley

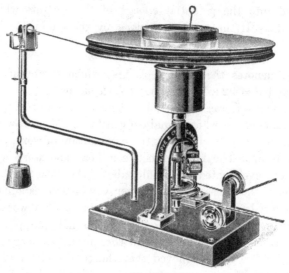

Fig. 22.

and a weight is hung from it. The inside part of the calorimeter contains water, the temperature of which is measured with a thermometer. A little oil should be put between the two conical vessels so that the inner one can turn round freely in the other. The outside vessel is made to rotate at such a speed that the couple exerted by it on the inner vessel is just enough to support the weight.

If r is the radius of the wheel in cms. the couple is then equal to rmg dyne-cms., where m is the mass of the weight in grams. The work done per revolution is therefore equal to $2\pi rmg$. If the outer vessel makes n revolutions the total work done is

$2\pi nrmg$. If the temperature of the water in the calorimeter rises from t_1 to t_2 we have

$$2\pi nrmg = J \; (w + w') \; (t_2 - t_1),$$

where w is the mass of the water in grams and w' the heat capacity of the calorimeter and thermometer. In this way J can be found within about 2 per cent. The temperature t_2 should be corrected for loss of heat during the experiment in the usual way.

Rowland made a series of very exact measurements of J at Baltimore, U.S.A., in the years 1877–78 by a method similar in principle to that just described. His calorimeter held about 9000 grams of water and its temperature could be raised 25° C. in 40 minutes. A steam engine was used to drive the apparatus. Rowland found that the energy required to raise one gram of water one degree centigrade was not exactly the same for different degrees of temperature. The following table gives some of Rowland's results:

Temperature	Energy required to raise a gram of water 1° C.
6° C.	$4\cdot203 \times 10^7$
10	$4\cdot196$ „
20	$4\cdot181$ „
30	$4\cdot174$ „

Subsequent researches have shown Rowland's results to be nearly exact. The number of ergs required to raise one gram of water from 15° C. to 16° C. is very nearly $4\cdot19 \times 10^7$. The number of foot-pounds of work required to raise the temperature of a pound of water one degree Fahrenheit is about 778.

CHAPTER X

THE CONVERSION OF HEAT INTO WORK

IF a gas is compressed it gets hotter. This may be shown by means of what is called a fire syringe. A fire syringe is merely a small brass cylinder with a piston and piston rod. If a piece of tinder is put in the cylinder and the piston quickly forced down so as suddenly to compress the air in the cylinder to a small volume, the air gets so hot that the tinder catches fire. When a gas is compressed work is done on it which is converted into heat. If a gas is allowed to expand so that it does work on the walls of the vessel containing it the gas gets colder, and heat must be imparted to the gas to bring it back to its original temperature. The amount of heat required to raise the temperature of a gas therefore depends on whether the volume is kept constant or is allowed to change when the gas is heated. If the volume is kept constant the gas does no work, so that the heat required is used up in raising the temperature of the gas. The amount of heat required to raise the temperature of a unit mass of gas one degree when its volume is kept constant is called its specific heat at constant volume and will be denoted by C_v. The specific heat at constant volume can be found by first measuring the heat capacity of a vessel filled with the gas and then the heat capacity of the same vessel when empty. Joly measured the specific heat at constant volume of several gases in this way. The heat capacity was found by immersing the vessel in steam and finding the mass of water which condensed on it while its temperature rose to the temperature of the steam. If m grams of water condense, then

$$mL = (w' + GC_v)(t_2 - t_1),$$

where L is the latent heat of vaporization of water at the temperature t_2 of the steam, w' the heat capacity of the vessel, G the mass of the gas and t_1 the initial temperature of the vessel. The following table gives some values of C_v in calories per gram, at pressures not above one atmosphere, as found by Joly:

Gas	C_v
Air	0·172
Hydrogen	2·40
Carbon dioxide	0·165

The specific heats of gases at constant volume are difficult to measure accurately, so that they are usually found indirectly by calculation from the specific heats at constant pressure and the ratio of the specific heat at constant pressure to that at constant volume. This ratio can be found by methods which will be described later.

If a gas is heated and allowed to expand so that its pressure remains constant, then the heat required to raise the temperature of unit mass one degree is called its specific heat at constant pressure. It will be denoted by C_p.

The specific heat at constant pressure can be found by passing a stream of gas, heated to a known temperature, through a spiral tube immersed in water in a calorimeter. The gas is cooled down to the temperature of the water and the heat it gives out can be found from the rise in the temperature of the water in the usual way. The following table gives some values of C_p in calories per gram:

Gas	C_p
Air	0·2389
Hydrogen	3·410
Carbon dioxide	0·1952

The ratio C_p/C_v can be deduced from the velocity of sound in the gas, as we shall see later. The following table gives the values of C_p/C_v and also of C_v calculated from the values of C_p by means of the ratio C_p/C_v:

Gas	C_p/C_v	C_v
Air	1·41	0·170
Hydrogen	1·41	2·42
Carbon dioxide	1·30	0·150

It will be seen that the values of C_v got in this way differ slightly from those found by Joly.

Let AB (Fig. 23) be a circular hollow cylinder closed at B, and let CD be a piston which can slide freely along the cylinder. Suppose that the space below the piston is filled with a gas such as air or hydrogen. Let the pressure of the gas be denoted by p and its temperature by t_1. Suppose that the cylinder is heated and that as the temperature rises the piston moves up so that the pressure of the gas is kept constant. If A denotes the area of cross section of the cylinder the force on the piston is Ap, and if the piston moves up a distance d the work done by the force is Apd. But Ad is the increase in the volume of the gas, so that if the volume increases from v_1 to v_2 the work done by the gas on the piston is $p(v_2 - v_1)$.

Mayer's Calculation of J.

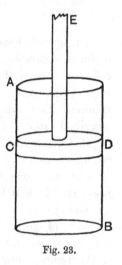

Fig. 23.

Let the temperature rise from t_1 to t_2 when the volume of the gas increases from v_1 to v_2 at constant pressure. Then we have

$$v_1 = v_0 (1 + \alpha t_1),$$
$$v_2 = v_0 (1 + \alpha t_2),$$

where v_0 is the volume of the gas at $0°$ C. and α its coefficient of expansion at constant pressure. Hence $v_2 - v_1 = v_0 \alpha (t_2 - t_1)$, so that W the work done is given by $W = p v_0 \alpha (t_2 - t_1)$.

When the gas is heated at constant volume it does no work, but when it is heated at constant pressure it does the work just calculated. In 1842, before Joule's researches, Mayer suggested that the greater amount of heat required to heat a gas at constant pressure is necessary because some of the heat is used up in doing the work. Suppose the mass of the gas is m, then the difference between the heat required at constant pressure and that at constant volume is equal to $m(C_p - C_v)(t_2 - t_1)$. If J denotes the amount of work equivalent to one unit of heat, then Mayer supposed that

$$p v_0 \alpha (t_2 - t_1) = J m (C_p - C_v)(t_2 - t_1).$$

Thus Mayer supposed that when a gas is heated at constant

pressure the heat actually remaining in the gas is the same as at constant volume and that the additional heat required at constant pressure does not stay in the gas but is used up in doing the work the gas does when it expands. The equation just obtained by making this important assumption reduces to

$$J = \frac{p v_0 \alpha}{m (C_p - C_v)} = \frac{p \alpha}{\rho_0 (C_p - C_v)},$$

where $\rho_0 = \dfrac{m}{v_0}$ is the density of the gas at $0°$ C. and at the pressure p. For air we have $C_p = 0\cdot2389$ and $C_v = 0\cdot1700$ calories per gram. Also if $p = 760$ mm., $\rho_0 = 0\cdot001293$; and $\alpha = 0\cdot00367$.

Hence
$$J = \frac{76 \times 13\cdot596 \times 980\cdot6 \times 0\cdot00367}{0\cdot001293 \times 0\cdot0689}$$

$$= 4\cdot18 \times 10^7 \text{ ergs per calorie.}$$

This result agrees very well with the value of J found by direct experiment, which shows that Mayer's assumption was nearly correct. According to Mayer's assumption we should expect that if the gas were allowed to expand without doing any work then the heat required to raise its temperature from t_1 to t_2 would be the same as at constant volume. We may suppose that the gas first expands from v_1 to v_2 without doing any work and without receiving any heat and is then heated at constant volume to t_2. It is found that the specific heat at constant volume is nearly the same whatever the volume, so that, according to Mayer's assumption, the first operation of expanding from v_1 to v_2 without doing any work ought to leave the gas at its initial temperature t_1, and thus the heat required for the final operation would be equal to $m C_v (t_2 - t_1)$, the same as for merely heating from t_1 to t_2 at constant volume. According to Mayer's assumption we should therefore expect that on allowing a gas to expand without doing any work its temperature would not change.

Gay-Lussac and, later, Joule tried experiments to test this point. Two large vessels of about equal volumes were connected by a pipe containing a cock. One vessel was exhausted and the other contained air at a high pressure. On opening the cock the air expanded to twice its volume without doing any work on the whole and it was found that when the whole apparatus was immersed in water in a calorimeter no appreciable change of temperature took place. More exact experiments made later by a different method have shown that there is really a slight cooling effect when air expands without doing any external work. This effect however is very small so that it appears that Mayer's assumption is almost exactly true.

When a gas is compressed it gets hotter so that to keep it at a constant temperature heat must be removed from it. It follows from the truth of Mayer's assumption that this heat must be an amount of energy equal to the work done on the gas in compressing it. If no heat is removed from the gas then, since it gets hotter, it follows that its pressure increases more rapidly as its volume is diminished than it would if the temperature were kept constant. When the state of a substance is changed in any way and no heat is allowed to enter or leave it during the change, the change is called an *adiabatic* change. If the temperature is kept constant during a change, then the change is called an *isothermal* change.

Indicator Diagrams.

It is difficult in practice to prevent heat from entering or leaving a substance when its volume and pressure are changed.

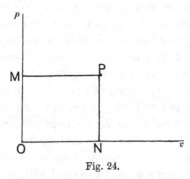

Fig. 24.

We may however imagine the substance to be enclosed in a vessel the walls of which are perfect non-conductors of heat, and we can discuss theoretically the properties a substance would have under such conditions. In practice, if the change in the state of the substance is made very rapidly, there may not be time for an appreciable amount of heat to enter or leave the substance, so that very rapid changes of state are often practically adiabatic changes.

For a gas like air or hydrogen we have the equation

$$p_t v_t = p_0 v_0 \left(1 + \alpha t \right).$$

If the temperature t is kept constant, then pv is constant. We may represent the relation between the pressure and the volume of a definite mass of a gas by means of a diagram in which distances measured vertically upwards from a horizontal line represent the pressures, and distances horizontally from a vertical line represent the volumes. Such a diagram is shown in Fig. 24. The point P corresponds to the state of the gas when its pressure is represented by NP and its volume by MP or ON. If we suppose the gas consists of one gram of air, then when its pressure is 76 cms. of mercury and its temperature $0°$ C. its volume is $\dfrac{1}{0\cdot 001293} = 773\cdot 4$ c.c.

In Fig. 25 the curve marked $t = 0°$ C. represents the relation between p and v for one gram of air at $0°$ C. The pressure p is expressed in cms. of mercury and the volume in cubic cms. We have $p_t v_t = p_0 v_0 \left(1 + \alpha t \right)$ or, since $\alpha = \dfrac{1}{273}$, $p_t v_t = p_0 v_0 \left(\dfrac{273 + t}{273} \right)$. If we take $t = 273°$ C. it follows that $p_t v_t$ at $273°$ C. is double $p_0 v_0$. The curve marked $t = 273°$ C. shows the relation between p and v for one gram of air at $273°$ C. The other two curves show the same thing at temperatures $2 \times 273 = 546°$ C. and $3 \times 273 = 819°$ C. Along each curve pv is constant and for each curve it is proportional to $273 + t$, so that, for the curves drawn, the values of pv are proportional to 1, 2, 3 and 4. These curves are called isothermal curves because the temperature corresponding to points on each is constant. Along a horizontal line like $ABCDE$ the distances AB, AC, AD and AE represent the volumes of one gram of air at the constant pressure represented by OA and the temperatures $0°$, $273°$, $546°$ and $819°$ C. These volumes are proportional to $273 + t$, that is to 1, 2, 3 and 4. Along a vertical line like $KHGFE$ the distances KH, KG, KF and KE represent the pressures of one gram of air at the constant volume represented by OK and the temperatures $0°$, $273°$, $546°$ and $819°$ C. These pressures are proportional to $273 + t$, that is to 1, 2, 3 and 4. Similar isothermal curves may of course be drawn for any other temperatures.

Suppose the pressure and the volume of the gas are represented
by the point M and that the gas is enclosed in a cylinder and piston
made of a perfect non-conductor of heat. If the piston is forced
down so as to diminish the volume of the gas it will get hotter so
that its pressure will rise more rapidly than if its temperature
were kept constant. The relation between the pressure and

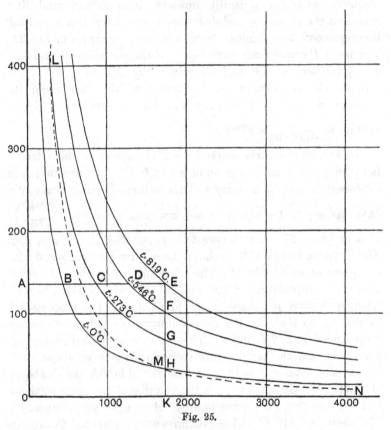

Fig. 25.

volume will therefore be represented by a curve like the dotted
line LMN which is steeper than the isothermal curves. A curve
like LMN which represents the relation between the pressure
and the volume when no heat enters or leaves the gas is called an
adiabatic curve. The temperature corresponding to any point on
an adiabatic curve is the temperature of the isothermal curve

passing through the same point. Thus the temperature corresponding to L is $273°$ C., and to M it is $0°$.

The isothermal curves shown in Fig. 25 also represent the relation between the pressure and volume of any other gas for which $p_t v_t = p_0 v_0 (1 + \alpha t)$, provided the volume of the other gas taken is equal to that of one gram of air at the same pressure and temperature.

A diagram showing the relations between the pressure and the volume of any body is called an *indicator diagram*. In Fig. 26 let

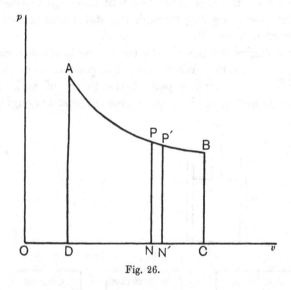

Fig. 26.

AB be a curve showing the relation between p and v for some quantity of any substance. If the substance changes from the state represented by P to that represented by a point P' close to P, then the increase of volume is represented by NN'. If P and P' are very close together the pressure at P' will be practically equal to that at P so that, since the work done by the substance when it expands from P to P' is equal to $p\,(v_2 - v_1)$ where p is its pressure and $v_2 - v_1$ the increase of volume, the work done by the substance in going from P to P' will be represented by $PN \times NN'$, that is by the area $PNN'P'$. If on the diagram p is expressed in dynes per sq. cm. and the volume in c.c. the area will represent the work done in ergs. If the pressure is in pounds weight per square foot and the volume in cubic feet, the area will represent the work done in foot-pounds. If the substance goes from the state represented by A to the state represented by B along the curve $APP'B$, then the work done by the substance will be represented by the area $ABCD$. For the whole curve AB can be divided into a great many short bits like PP' and in the same way as for PP' the work corresponding to each bit is represented by the area of the vertical strip below it like $PNN'P'$. But all the strips make up the area $ABCD$ so that this area represents the total work done in going through the states represented by the curve between A and B.

A heat engine is a machine for converting heat into mechanical work. We are now in a position to consider the elementary part of the theory of such engines. The most important kind of heat engine is the steam engine. We

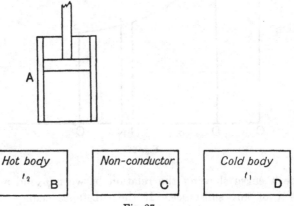

Fig. 27.

shall first consider a simple form of heat engine the theory of which can be easily worked out. This engine is called Carnot's ideal heat engine. It works on the same general principles as real engines but is much simpler. It cannot be realised in practice, but in real heat engines an attempt is made to approximate to the theoretically perfect conditions under which Carnot's ideal engine is imagined to work. Fig. 27 shows Carnot's engine. A is a cylinder and piston. The side walls of the cylinder and the piston are supposed to be made of material which is a perfect non-conductor of heat. The bottom of the cylinder is supposed to be made of a perfect conductor of heat. The cylinder contains a quantity of some substance such as a gas, like air, or a liquid, like water, and its vapour. The gas or the vapour of the liquid exerts a pressure on the piston and walls of the cylinder. The piston is supposed to be held down by a force equal to the upward force on it due to this pressure. B is a hot body which is a perfect conductor of heat and is supposed to be maintained at a constant temperature t_2 by means of a source of heat. C is a non-conducting block. D is a cold body which is a perfect conductor of heat and is supposed to be maintained at a constant temperature t_1.

If the cylinder contains water and water vapour the pressure in it will be equal to the vapour pressure of the water, and will remain constant when the piston is moved up or down, provided the temperature is kept constant and both water and water vapour are present. If the piston is pushed down till all the vapour is condensed, then to push it down much further requires a very great force because water is almost incompressible. Also if the piston is raised until all the water has evaporated into saturated vapour, then on raising it still further the pressure will get less because the vapour will then no longer be saturated.

Let the relation between the pressure p in the cylinder and the volume v enclosed beneath the piston be represented on the indicator diagram shown in Fig. 28. Suppose that at the start the temperature of the substance in the cylinder is t_2 and that the bottom of the cylinder is put on the top of the hot body B. Let the pressure and volume at the start be represented by the point A on the indicator diagram.

Suppose now that the piston is allowed to rise very slowly so that the volume v increases. The substance then does work on the piston so that its temperature tends to fall, but since the bottom of the cylinder which is a perfect conductor is on the hot body at temperature t_2, heat will flow into the substance from the hot body and prevent its temperature falling appreciably. We suppose the piston rises so slowly that the temperature of the substance remains constant at t_2. If the piston were allowed to rise quickly heat would not be able to get into the substance fast enough from the hot body to keep the temperature of the whole of the substance from falling below t_2.

Let the piston rise slowly till the volume and pressure are represented by B, and let AB represent the relation between the pressure and volume during the expansion, so that AB is an isothermal curve for the temperature t_2. Let the amount of heat absorbed from the hot body during this operation be denoted by H_2.

If the cylinder contains water and water vapour AB will be a horizontal line, and the heat absorbed will be the latent heat of

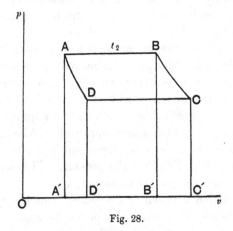

Fig. 28.

evaporation of the water which evaporates when the piston rises. Now let the cylinder and piston be removed from the hot body B and put on the non-conductor C. The substance in the cylinder is then completely surrounded by non-conductors of heat so that no heat can enter or leave it. Let the piston now be allowed to again rise slowly. The substance then does more work on the piston and its temperature falls. Let the expansion continue till the temperature has gone down to t_1, that of the cold body D.

Suppose BC represents this expansion. BC is then an adiabatic curve. The cylinder and piston are now put on the cold body and the piston is pushed down very slowly. Work is then done on the substance and it tends to get hotter, but if the motion of the piston is slow enough the cold body will remove heat from it fast enough to prevent its temperature from rising above t_1. Let CD represent this isothermal compression at t_1. The compression is stopped when the point D which is on the adiabatic curve through A is reached. Let the amount of heat given up to the cold body be denoted by H_1. The cylinder and piston are now again put on the non-conductor C and the piston is very slowly forced down until the temperature of the substance has risen to its original value t_2. This adiabatic compression is represented by DA because D was chosen so as to be on the adiabatic curve through A. The substance has now been brought back exactly to its original state represented by the point A, so that the amount of energy in it must be the same as at the start. The work done by the substance during the expansion along ABC is represented by the area $ABCC'A'$. The work done on the substance during the compression along CDA is represented by the area $CDAA'C'$. The difference between the work done by the substance on the piston and the work done on the substance by the piston is therefore represented by the area $ABCD$. The series of operations described is called a Carnot's cycle because it was invented by Carnot and brings the substance back to its initial state. The result of the cycle is that a total amount of work represented by the area $ABCD$ has been done by the substance and an amount of heat H_2 has been taken from the hot body and an amount of heat H_1 has been given to the cold body. Let W denote the work represented by $ABCD$. An amount of work W has been done and an amount of heat $H_2 - H_1$ has disappeared. We therefore conclude that

$$W = J\,(H_2 - H_1),$$

where J denotes as usual the mechanical equivalent of heat. The efficiency of a heat engine may be defined to be the amount of work it does divided by the mechanical equivalent of the heat which it receives from the source of heat used to drive it. The efficiency E of Carnot's engine is therefore given by the equation

$$E = \frac{W}{JH_2}.$$

Let us now consider what would happen if we worked the engine backwards. Start at A and expand adiabatically to D

and then isothermally on the cold body to C; then compress adiabatically to B and isothermally on the hot body back to A. If these operations were done extremely slowly the substance would go backwards round exactly the same cycle of pressures and volumes represented by $ADCB$ as it previously went round forwards.

If the cycles were not done very slowly the pressures and volumes could not be the same when going backwards as when going round forwards. Consider the expansion from A to B. If this were done quickly the temperature would fall below t_2 during it, so that the actual relation between the pressure and the volume would be a curve joining the points A and B but lying below the isothermal curve AB. During the backwards compression from B to A the temperature would rise above t_2 so that the actual relation between the pressure and the volume would be represented by a curve joining B to A but lying above the isothermal curve BA. Only if the cycle is gone through extremely slowly can it be represented by the same curves whether done backwards or forwards. A cycle which is gone through in such a way that it can be represented on the indicator diagram by the same curves whether gone through backwards or forwards is called a reversible cycle. Carnot's engine is supposed to work in a reversible cycle, and it is therefore called a perfectly reversible engine. The result of the cycle when done backwards is that a quantity of heat H_1 has been taken from the cold body, a quantity H_2 given to the hot body and a total amount of work W represented by the area $ABCD$ has been done on the substance. As before $W = J(H_2 - H_1)$.

It is now necessary to consider a principle called the *second law of thermo-dynamics* which is based on experience. It is found that heat does not pass from cold to hot bodies by itself. It can only be made to do so by the expenditure of work. Heat flows by conduction from hot to cold bodies but never from cold to hot bodies. When Carnot's engine is worked backwards heat is taken from the cold body and more heat is given to the hot body, but an amount of work W has to be done. If we could devise a process for making heat go from cold to hot bodies without the expenditure of work we should be able to concentrate heat by means of it into any desired body and so should not need to burn coal or other fuel to get high temperatures. Nothing in our experience justifies us in supposing that anything of the sort is possible.

By means of this principle we can show that no engine using the same hot and cold bodies can be more efficient than a perfectly reversible engine, and also that the efficiency of a perfectly reversible engine is independent of the nature of the substance used in its cylinder and depends only on the temperatures t_2 and t_1 of its hot and cold bodies. Suppose, if possible, that there is another kind of heat engine which can take heat from the hot body B at t_2, convert more of this heat into work than Carnot's engine and give the rest of the heat to the cold body D at t_1.

Let this engine be used to drive Carnot's engine backwards so that all the work the more efficient engine does is used up in driving Carnot's engine. For the Carnot's engine we have

$$W = J\,(H_2 - H_1) \text{ and } E = W/JH_2,$$

for the other engine suppose

$$W = J\,(H_2' - H_1') \text{ and } E' = W/JH_2'.$$

W is the same for both engines.

Here H_2' is the heat the other engine takes from the hot body and H_1' is the heat it gives to the cold body. E' the efficiency of the second engine we suppose is greater than E. If $E' > E$, then H_2 must be greater than H_2'. But $W = J\,(H_2 - H_1) = J\,(H_2' - H_1')$; hence, since $H_2 > H_2'$ we have $H_1 > H_1'$.

As the result of the action of the two engines the hot body gains heat H_2 from the Carnot's engine and loses heat H_2' to the other engine. On the whole then the hot body gains an amount of heat $H_2 - H_2'$, which is positive because $H_2 > H_2'$. The cold body in the same way loses an amount of heat $H_1 - H_1'$ equal to $H_2 - H_2'$. The two engines therefore convey heat from the cold body to the hot body without any work being done on them from outside. This is contrary to experience, and according to the second law of thermo-dynamics is impossible. Hence E' cannot be greater than E; so no engine can be more efficient than Carnot's engine.

If possible let another perfectly reversible engine using the same hot and cold bodies be *less* efficient than the Carnot's engine. Then, if the Carnot's engine is used to drive this less efficient but also perfectly reversible engine backwards, heat will be conveyed from the cold body to the hot body without any work being done, so that it is impossible for any perfectly reversible engine using the same hot and cold bodies to be less efficient than Carnot's engine. Hence all perfectly reversible engines using the same hot and cold bodies or working between the same temperatures t_2 and t_1 have equal efficiencies. The efficiency of a perfectly reversible engine is therefore independent of the nature of the working substance in it and depends only on the temperatures between which it works. If the efficiency of a perfectly reversible engine working between the temperatures t_2 and t_1 is E, then we have

$$E = f(t_1, t_2),$$

where $f(t_1, t_2)$ denotes some function of t_1 and t_2 only.

We have then in E a quantity depending only on temperatures and not on the special properties of particular substances. The scale of temperature given by the constant volume hydrogen thermometer is an arbitrary scale, for it depends on the properties of hydrogen. The scale depending on the expansion of a liquid like mercury is equally arbitrary. Lord Kelvin proposed to define a scale of temperature by means of the relation $E = f(t_1, t_2)$, and so to get a scale of temperature independent of the properties of any particular substance.

To do this all that is necessary is to choose a particular form for the function f. If $t_1 = t_2$ it is easy to see that E must equal zero, for then the two isothermal curves in the Carnot cycle coincide, so that the area $ABCD$ which represents W is zero. This condition is satisfied if $f(t_1, t_2) = A(t_2 - t_1)$, where A is a quantity to be determined.

Suppose then that

$$E = A(t_2 - t_1).$$

If E were equal to unity all the heat taken from the hot body would be converted into work. Let the temperature of the cold body for which $E = 1$ be taken as the zero of the new scale of temperature. We then have $t_1 = 0$ when $E = 1$ so that

$$E = 1 = At_2, \text{ or } A = \frac{1}{t_2}.$$

Hence $$E = \frac{t_2 - t_1}{t_2}.$$

This relation defines a new scale of temperature which is called the absolute scale, because it does not depend on the properties of any particular substance.

The size of the degrees on this new scale are chosen so that there are 100 degrees between the freezing point of ice and the boiling point of water at 76 cms. of mercury pressure. We might, if it were possible, determine the efficiency of a perfectly reversible engine working between these two temperatures; it would be found to be equal to nearly 0·2681. We should then have

$$0·2681 = \frac{100}{t_2}.$$

Hence $t_2 = 373$ and therefore $t_1 = 273$. Thus the freezing point of ice is at $273°$ on the absolute scale and the boiling point of water at $373°$ when the interval between these two temperatures is made equal to $100°$. Practically it is not possible to construct a perfectly reversible engine, so that the absolute scale cannot be realised in this way. However we can show that it must coincide with the scale of temperature given by a constant volume gas thermometer containing an ideal gas the temperature of which remains constant when it expands without doing work. Hydrogen gas very nearly satisfies this condition, so that it follows that the scale of temperature given by the standard hydrogen thermometer is nearly the same as the absolute scale.

Fig. 29 shows an indicator diagram for a Carnot engine containing the ideal gas just mentioned. When expanding isothermally from A to B the work done by the gas is represented by the area $ABB'A'$. But for such an ideal gas the heat absorbed is mechanically equivalent to this work. Hence JH_2 is represented by the area $ABB'A'$. The efficiency of the engine is therefore given by

$$E = \frac{W}{JH_2} = \frac{\text{area } ABCD}{\text{area } ABB'A'}.$$

If AB and CD are very near together, that is, if t_2 and t_1 are nearly equal, we have

$$\frac{\text{area } ABCD}{\text{area } ABB'A'} = \frac{PQ}{PR} = \frac{PR - QR}{PR}.$$

Hence
$$E = \frac{t_2 - t_1}{t_2} = \frac{PR - QR}{PR}.$$

But PR and QR represent the pressures of the gas at the temperatures t_2 and t_1 and the constant volume OR. It appears therefore that the pressure of the gas at constant volume is proportional to the absolute temperature.

The pressure p of the hydrogen in the standard hydrogen thermometer is therefore very nearly proportional to the absolute temperature. At the melting point of ice $p = 100$ cms. of mercury, and at the boiling point of water $p = 136\cdot67$ cms. of mercury.

If t_0 denotes the melting point of ice on the absolute scale and if we take the boiling point of water equal to $t_0 + 100$ so as to

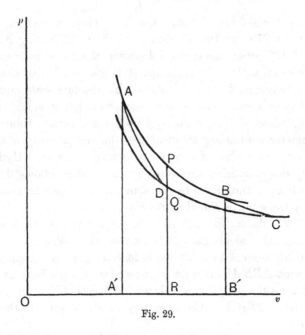

Fig. 29.

have 100 degrees between the melting point of ice and the boiling point of water, we have

$$\frac{t_0 + 100}{t_0} = \frac{136\cdot67}{100},$$

which gives $t_0 = 272\cdot7°$ or nearly $273°$. Hence $0°$ C. is equal to nearly $273°$ on the absolute scale, and so $t°$ C. is equal to nearly $t + 273°$ on the absolute scale.

The equation $p_t v_t = p_0 v_0 (1 + \alpha t)$, where t denotes temperature on the centigrade scale, since $\alpha = \frac{1}{273}$ nearly, may be written

$$p_t v_t = p_0 v_0 \left(\frac{273 + t}{273} \right).$$

Let absolute temperature now be denoted by θ so that $\theta = 273 + t$ and we get

$$pv = \frac{p_0 v_0 \theta}{273} = R\theta,$$

where R is a constant equal to $p_0 v_0 / 273$. Thus for any gas the product pv is nearly proportional to the absolute temperature so long as the gas approximately obeys the laws expressed by the equation

$$pv = p_0 v_0 \left(1 + \frac{t}{273} \right).$$

Since the density of any gas is nearly proportional to its molecular weight it follows that the volume of a number of grams of the gas equal to its molecular weight is nearly the same for all gases. The volume of 32 grams of oxygen is equal to $\frac{32}{0 \cdot 0014292} = 22400$ c.c. at $0°$ C. and 76 cms. of mercury pressure. The value of the constant $R = p_0 v_0 / 273$ for a molecular weight in grams of any gas is therefore nearly equal to

$$\frac{1 \cdot 014 \times 10^6 \times 22400}{273} = 8 \cdot 32 \times 10^7 \text{ ergs.}$$

R is expressed in ergs because p_0 is expressed in dynes per sq. cm. and v_0 in cubic cms. This value of R is nearly equal to two calories of heat since one calorie is equal to $4 \cdot 19 \times 10^7$ ergs. It is often convenient to put $R = 2$ calories for one molecular weight in grams of any gas. In the equation $pv = 2\theta$ the unit of energy is the calorie so that the unit of pressure is a pressure of $4 \cdot 19 \times 10^7$ dynes per sq. cm., which is equal to $31 \cdot 4$ metres of mercury.

The equation $J (C_p - C_v) m = p v_0 \alpha$ (see page 200), in which v_0 denotes the volume of a mass m of the gas at $0°$ C. and pressure p, since $R = p_0 v_0 \alpha$ and $pv_0 = p_0 v_0$ gives

$$J (C_p - C_v) m = R.$$

If m is taken equal to the molecular weight of the gas in grams, this becomes

$$(C_p - C_v)\, m = 2 \text{ calories.}$$

The following table gives some values of C_p, C_v, m, and of $(C_p - C_v)\, m$:

Gas	C_p	C_v	m	$(C_p - C_v)\, m$
Air	0·2389	0·170	28·74	1·98
Hydrogen	3·41	2·42	2·016	2·00
Carbon dioxide	0·1952	0·150	44·00	1·99

It will be seen that $(C_p - C_v)\, m$ is very nearly equal to 2 calories for these gases.

We have seen that the efficiency of a perfectly reversible engine is equal to $(t_2 - t_1)/t_2$, where t_2 and t_1 are **Efficiency of Heat Engines.** the temperatures between which it works on the absolute scale. To get a high efficiency it is clear that the hot body or source of heat should be as hot as possible and the cold body as cold as possible. In actual heat engines, such as steam engines, the condenser where the steam is condensed after doing work in the engine corresponds to the cold body, and the hot water in the boiler may be taken to correspond to the hot body. The condenser cannot be kept below about 15° C. to 20° C., so that to increase the theoretically possible efficiency of a steam engine the only practical plan is to raise the temperature of the water in the boiler. The higher the temperature of the water the higher the pressure of the steam. It is difficult to make large boilers which will work safely at pressures much above 400 pounds weight per sq. inch. The temperature of the water at which its vapour pressure is equal to 400 pounds per sq. inch is about 230° C. The theoretically possible efficiency of a heat engine working between 230° C. and 15° C. is equal to

$$\frac{(230 + 273) - (15 + 273)}{230 + 273} = 0\cdot 43.$$

Such a heat engine therefore cannot convert into work more than 43 % of the heat which gets into the water in the boiler. At least 57 % of this heat must be given up to the condenser and cannot be converted into work. In practice this theoretically possible efficiency cannot be obtained. The temperature of the furnace of an engine may be 1500° C. If the engine could be made to work between this temperature and 15° C. its possible efficiency would be

$$\frac{1500 - 15}{1500 + 273} = 0\cdot84$$

or about 84 %.

REFERENCES

Theory of Heat, Clerk-Maxwell.
Heat, Poynting and J. J. Thomson.

CHAPTER XI

THE KINETIC THEORY OF GASES

THE chief properties of gases can be very completely explained
by the theory according to which a gas consists of an immense
number of independent molecules moving about in the space
occupied by the gas. This theory is called the kinetic theory of
gases. In a gas at 0°C. and 76 cms. of mercury pressure there are
about $2·6 \times 10^{19}$ molecules per cubic centimetre. The "diameter"
of each molecule is about 3×10^{-8} cms., so that the total volume
of all the molecules in one c.c. is only about

$$2·6 \times 10^{19} \times 14 \times 10^{-24} = 3·6 \times 10^{-4} \text{ c.c.}$$

Thus only about four parts in ten thousand of the total volume is
actually occupied by the molecules in a gas at 0°C. and 76 cms.
pressure. The molecules collide with each other and with the
walls of the vessel containing the gas. The collisions between
the molecules continually change their velocities in direction and
in magnitude and produce a certain average distribution of
velocities among the molecules. The pressure of the gas on the
walls of the vessel containing it is due to the impacts of the
molecules. The molecules are supposed to be perfectly elastic, so
that when they hit the walls of the vessel they bounce off
without loss of velocity.

Consider a plane area of one square cm. on the wall of the
vessel containing the gas. Suppose the gas near this area
contains n molecules per c.c., each of mass m, that have velocity
components towards the area equal to v. The number of these
molecules striking the area per second will be nv. Each of these
molecules when it strikes the surface and bounces off will com-
municate an amount of momentum $2mv$ to the surface, for the
momentum of the molecule in the direction towards the surface

is changed by the impact from mv to $-mv$. The total momentum received by the sq. cm. of surface per second due to these molecules is therefore equal to $nv \times 2mv$ or $2nmv^2$. We may suppose all the molecules moving towards the area divided up into groups like the group just considered, and for each group the momentum communicated to the sq. cm. of area will be equal to twice the mass per c.c. of the group multiplied by the square of its velocity component towards the area. Half the molecules will be moving towards the area and half away from it. Consequently if ρ is the density of the gas and $\bar{v}^2$ denotes the average value of the squares of the velocity components normal to the sq. cm., the total momentum gained by the sq. cm. of area per second is $\rho\bar{v}^2$. Each molecule at any instant has a velocity U which can be resolved into three perpendicular components u, v and w such that $U^2 = u^2 + v^2 + w^2$. Since on the average as many molecules move one way as another it follows that the average value of u^2 is equal to that of v^2 and of w^2. Hence $\bar{U}^2 = 3\bar{v}^2$, where $\bar{U}^2$ denotes the average value of U^2. The momentum gained per sec. by the sq. cm. is therefore equal to $\tfrac{1}{3}\rho\bar{U}^2$. But since a force is equal to the momentum it imparts per sec. to a body, it follows that

$$p = \tfrac{1}{3}\rho\bar{U}^2,$$

where p denotes the pressure or force per unit area exerted by the gas on the walls of the vessel containing it. Let V denote the volume of one gram of the gas; then $\rho = 1/V$ and $pV = \tfrac{1}{3}\bar{U}^2$. This equation shows that so long as $\bar{U}^2$ remains constant, that is, so long as the average kinetic energy of translation of the molecules is constant, the product pV remains constant. In a gas at constant temperature we may suppose that the average kinetic energy of the molecules is constant, so that the result just obtained is in agreement with Boyle's law. If we suppose that the average kinetic energy of the molecules is proportional to the absolute temperature, then the equation $pV = \tfrac{1}{3}\bar{U}^2$ shows that pV is proportional to the absolute temperature. This agrees with the equation $pV = R\theta$ previously obtained. Thus the kinetic theory explains in a simple way the chief properties of gases.

By means of the equation $p = \tfrac{1}{3}\rho\bar{U}^2$ we can calculate the value of $\bar{U}$, that is of the square root of the average value of the squares of the velocities of the molecules. The following table gives the values of ρ and $\overline{U}$ for several gases at 0°C. and 76 cms. of mercury pressure calculated in this way. To calculate $\overline{U}$ it is necessary to express p in dynes per sq. cm. and ρ in grams per c.c. 76 cms. of mercury $= 1·014 \times 10^6$ dynes/cm.²

Gas	Density at 0°C. and 76 cms. grams/cm.³	$\overline{U}$ cms./sec.
Hydrogen ...	0·00009	183800
Oxygen ...	0·001429	46130
Nitrogen ...	0·001254	49260
Carbon monoxide	0·001251	49320
Helium ...	0·0001787	130500

The kinetic energy of the molecules in one c.c. of a gas is equal to $\frac{1}{2}\rho \overline{U}^2$ or, since $p = \frac{1}{3}\rho \overline{U}^2$, the kinetic energy is equal to $3p/2$. This is the energy of the translational motion of the molecules only, and does not include any energy they may have due to rotation or other forms of internal motion.

When a gas is heated so that its temperature rises, the energy of the molecules is increased. The total energy required to raise the temperature of one c.c. of a gas at constant volume from $t_1°$ to $t_2°$C. is equal to $J\rho C_v (t_2 - t_1)$. The part of this which goes to increase the velocities of the molecules is equal to

$$\tfrac{3}{2}(p_2 - p_1) = \tfrac{3}{2}p_0\alpha (t_2 - t_1).$$

Now $J(C_p - C_v) = \dfrac{p\alpha}{\rho_0}$ (see page 200). In this equation we may put $p = p_0$ because p/ρ_0 is independent of p. Hence the energy required to increase the velocities of the molecules is equal to $\tfrac{3}{2}\rho_0 J (C_p - C_v)(t_2 - t_1)$. The ratio of this to the total energy required to heat the gas at constant volume is therefore

$$\frac{3}{2}\frac{C_p - C_v}{C_v} = \frac{3}{2}(\gamma - 1),$$

where $\gamma = C_p/C_v$. For gases like helium, argon and mercury vapour it is found that $\gamma = 1\tfrac{2}{3}$, so that $\tfrac{3}{2}(\gamma - 1) = 1$, which shows that all the energy required to raise the temperature of these gases goes to increase the velocities of the molecules. For air, hydrogen and nitrogen $\gamma = 1\tfrac{2}{5}$, so that $\tfrac{3}{2}(\gamma - 1) = \tfrac{3}{5}$. For these gases therefore $\tfrac{2}{5}$ of the energy required to heat them at constant volume goes to increase the internal energy of the molecules.

Consider a cylinder and piston containing a gas. Let the piston move down with a uniform velocity v so as to compress the gas. Suppose a molecule having an upward velocity u collides with the piston. The velocity of the molecule relative to the piston before the impact is $u + v$. After the impact its velocity will be directed downwards, but will still be $u + v$ relative to the piston. The velocity after the impact is therefore equal to $u + 2v$. Every molecule which collides with the piston has its velocity normal to the piston increased by $2v$ in the same way so that the motion of the piston increases the average velocity of the molecules, and therefore makes the gas hotter. In the same way if the piston moves up so that the gas expands, the velocities of the molecules are decreased and the gas gets colder. If the gas is allowed to expand into a vacuum the velocities of the molecules are not changed, so that the average temperature of the gas does not alter.

If the number of molecules in one c.c. of a gas is n and each one has a mass m, then ρ the density of the gas is equal to nm, hence

$$p = \tfrac{1}{3} nm\, \overline{U}^2.$$

If we have another gas at the same pressure and temperature containing n' molecules per c.c. each of mass m', then

$$\tfrac{1}{3} nm\, \overline{U}^2 = \tfrac{1}{3} n'm'\, \overline{U}'^2,$$

where $\overline{U}'^2$ denotes the average value of the squares of the velocities of the molecules of the second gas. If we suppose that the average kinetic energies of the molecules of different gases are all equal at the same temperature, then we have

$$\tfrac{1}{2} m\, \overline{U}^2 = \tfrac{1}{2} m'\, \overline{U}'^2.$$

With the previous equation this gives $n = n'$, so that if the assumption just made is true all gases should contain equal numbers of molecules per c.c. when they are all at the same pressure and temperature. Their densities should therefore be proportional to their molecular weights, which is found to be the case. This shows that all gaseous molecules have nearly equal average kinetic energies of translation at any given temperature.

REFERENCES

Heat, Poynting and J. J. Thomson. *Kinetic Theory of Gases*, J. H. Jeans.

PART III

SOUND

CHAPTER I

PRODUCTION AND VELOCITY OF SOUND

THE sensation of sound is produced by a disturbance transmitted from sounding bodies through the air to the ear. The study of the processes taking place in the ear and brain belongs to psychology and physiology, while the study of the process taking place outside the head belongs to the branch of physics known as *sound*. Sound is produced by rapidly vibrating bodies and by any sudden disturbance of the air. For example, when a gun is fired

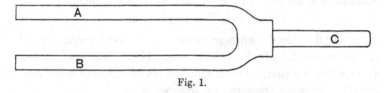

Fig. 1.

a large volume of gas is suddenly emitted by the gun, so that the surrounding air is violently pushed away from the muzzle of the gun. If one end of a short steel spring is held in a vice and the other end pulled sideways and let go, the spring vibrates rapidly backwards and forwards and gives out a sound. Fig. 1 shows a tuning fork, which is a symmetrical steel fork with two prongs *A* and *B* and a handle *C*. If the prongs are pressed towards each other and then let go, they vibrate and give out a sound. The fork can also be set vibrating by striking one of the prongs with a light wooden hammer covered with felt. The felt is intended to prevent damage to the fork. If a small ball made of ivory or any hard substance is hung up by a thread, then on touching it with one of the prongs of a sounding tuning fork the ball is violently knocked away from the fork, thus showing that the prongs are moving rapidly.

The sound emitted by a rapidly vibrating body is transmitted through the air to the ear. If the air surrounding the body is removed no sound is produced. This can be shown by hanging up a bell by a piece of india-rubber cord inside a closed vessel and pumping out the air with an air pump. If we shake the vessel and bell no sound is heard, but when we let in the air the bell can be heard distinctly outside the vessel. Sound can pass through any gas and also through solids and liquids. If a bell is let down into water and rung beneath the surface it can be heard above the water, or better by putting one's ear into the water. Soft inelastic solids like felt do not transmit sound readily. If a musical box is wrapped up in thick layers of felt it cannot be heard playing, but if the end of a long wooden or metal rod is passed through a small hole in the felt so as to touch the box and a wooden board is held against the other end of the rod, the music becomes distinctly audible. The sound from the box travels along the rod to the board and from the board escapes into the air. If the board is removed very little sound can be heard, for the surface of the rod is too small for much sound to escape from it. A vibrating tuning fork emits very little sound, but when the handle is held against a wooden board like a table top the sound becomes much louder. The fork sets the board vibrating and the sound escapes into the air from the board because of the larger surface.

Sound travels through air and other substances with a definite velocity that is different for different substances. When a gun is fired the flash is seen by an observer at a distance before the sound is heard. The velocity of light is so great that the time it takes to go from the gun to the observer can be neglected, and the time between seeing the flash and hearing the report can be taken to be equal to the time the sound takes to travel from the gun to the observer. The velocity of sound in air can be found in this way. If the distance between the gun and the observer is s and the time between seeing the flash and hearing the report is t, then

$v = s/t$, where v denotes the average velocity of the sound between the gun and the observer.

The velocity of sound in dry air at 0° C. is about 33200 cms./sec. or 1090 feet per second. Thus if the observer is 10900 feet away from the gun, the interval between the flash and the report is about 10 seconds. To find the velocity of sound in air accurately is very difficult. The velocity is affected by wind and depends on the temperature and humidity of the air. The distance used must be large to get a time interval which can be measured exactly, and it is difficult to get the temperature and humidity exactly over a large distance.

The velocity has been found fairly exactly by selecting two stations, at an exactly known distance, say 10 miles, apart, on hills with a valley in between. At each station a cannon is placed and an observer provided with a stop-watch or other form of chronograph to measure the time interval. Let the two stations be denoted by A and B. The observer at A fires his cannon and the observer at B measures the time between seeing the flash and hearing the report from A. The observer at B also fires his cannon as soon as he sees the flash at A and the observer at A measures the time between seeing the flash and hearing the report from B. The velocity is got by dividing the distance between A and B by the mean of the two times. The temperature and humidity are observed at A and B and possibly at intermediate points. If the wind is blowing from A to B with a velocity u, then the velocity of the sound from A to B will be $v + u$ and from B to A it will be $v - u$. Hence

$$t_1 = \frac{s}{v + u} \text{ and } t_2 = \frac{s}{v - u},$$

where t_1 is the interval measured at B, t_2 that measured at A and s the distance from A to B. The mean of t_1 and t_2 is therefore

$$t = \frac{t_1 + t_2}{2} = \frac{sv}{v^2 - u^2}.$$

If the velocity of the wind u is small, u^2 can be neglected compared with v^2, so that

$$t = \frac{s}{v} \text{ or } v = \frac{s}{t}.$$

If the wind is blowing across the direction from A to B then it carries the sound waves sideways, so that the sound takes longer to go either from A to B or from B to A than it would if there were no wind. In this case therefore the effect of the wind is not got rid of by taking the mean of the two time intervals. The best plan is to make measurements only on very calm days when there is little or no wind and when any wind there may be is blowing from A to B or from B to A. It is found that very powerful sounds travel quicker than ordinary sounds. Thus close to the cannon the velocity is greater than at a distance. This effect however is too small to produce an appreciable error when the velocity is measured over a distance of several miles. Except in the case of unusually intense sounds it is found that all sounds travel with the same velocity, nearly 33200 cms./sec. in dry air at 0° C. At a temperature $t°$ C. the velocity is equal to $33200 + 61t$ cms./sec. provided t is small. The velocity is found to be independent of the pressure of the air. It is the same at the sea level as between the tops of high mountains where the pressure is lower.

The velocity of sound in water has been found by striking a bell under the water in a large lake. A gun was fired above the bell by the same action that struck the bell. An observer at a great distance measured the time between seeing the flash of the gun and hearing the sound of the bell in the water. The velocity was found to be 143500 cms./sec. which is more than four times that in air.

REFERENCE

Sound, Poynting and J. J. Thomson.

CHAPTER II

WAVE MOTION

A DISTURBANCE produced at a point in a medium like air is propagated in all directions with a definite velocity. In a similar way a disturbance produced at a point on the surface of water is propagated in all directions over the surface. For example, if a small stone is thrown into a pond, waves spread out from the point where the stone enters the water in the form of circles the radii of which increase at a nearly uniform rate with the time. In such cases a movement travels through the medium, but after it has passed by, the medium is left in its original position or very near to it. The propagation of such movements through a medium is called wave motion. If we fix our attention on a particular particle in the medium, then this particle moves while the waves are passing over it but is left in or very near to its original position when the waves have gone beyond it. When the motion of the particle is parallel to the direction in which the waves are moving the waves are called longitudinal waves, and when the motion of the particle is in a plane perpendicular to the direction in which the waves are travelling the waves are called transverse waves. It is found that sound in air consists of longitudinal waves.

A powerful electric spark from a battery of Leyden jars produces a sound wave of great intensity. The spark very suddenly heats the air so that it expands and its pressure is suddenly increased. In this way an intense wave is started which moves outwards from the spark in the form of a thin hollow sphere with the spark at its centre. The radius of the sphere is equal to vt where v is the velocity of sound in air and t the time since the spark occurred. In the wave that is close to the surface of the sphere the density of the air is greater than in the surrounding space, and it is possible to take instantaneous photographs of the sound wave. This can be done by means of the light from

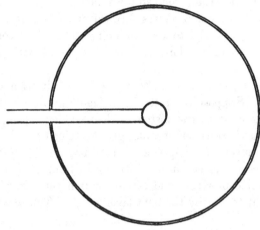

Fig. 2.

another electric spark. If the second spark is passed one ten-thousandth part of a second after the first one, the radius of the wave is about 3·3 cms. Fig. 2 represents a photograph taken in this way.

The way in which longitudinal waves are propagated can be illustrated with the model shown in Fig. 3. A number of solid lead spheres, each about 3 cms. in diameter, are hung up in a horizontal

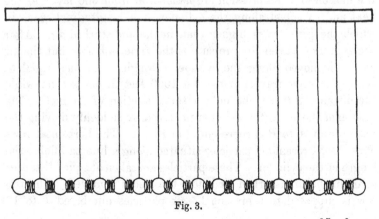

Fig. 3.

row by long cords. Between each sphere and the next a spiral spring is fixed. If the first sphere is suddenly pushed towards the next one, the spring is compressed and pushes the second sphere, and so compresses the second spring which moves the third sphere, and so on. Each sphere moves forward and compresses the next spring but is brought to rest in doing this. The compression can be seen to move along the row of spheres with a uniform velocity.

Consider a long cylinder CC, Fig. 4, full of air, with a piston P at one end. Suppose the piston is suddenly pushed in from A to B. The air close to the piston is then suddenly compressed so that its pressure rises, and it is also given a velocity in the direction AB. This layer of compressed air is brought to rest by the backward force exerted on it by the air in front of it, but at the same time it compresses and sets in motion this air, which is in turn brought to rest by the next layer of air. This next layer is

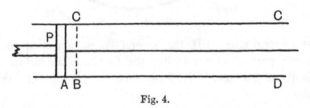

Fig. 4.

also set moving and compressed in the same way, and in turn passes on its motion and compression to the next layer, and so on. Thus the motion and compression are handed on from one layer to the next and so travel along the cylinder as a longitudinal wave in which the pressure is higher than in the undisturbed air. After the wave has passed the pressure is the same as before, but the air has been moved along the cylinder through a distance equal to AB. The wave travels more than 1000 feet in one second while the distance AB may be only a small fraction of an inch. The motion of the air in the cylinder produced by suddenly moving the piston from A to B is represented in Fig. 5. The horizontal rows of circles represent air particles situated along a line parallel to the length of the cylinder. These particles are taken at equal distances apart in the undisturbed air before the piston is moved. The top row is supposed to represent 17 air particles numbered 1 to 17 with the surface of the piston at A at the instant when the piston begins to move. The second row represents the piston and the same 17 particles at a short time τ after the piston begins to move. The piston has moved forward but the particles shown have not yet moved. The following rows show the positions of the particles and the piston at the ends of successive equal

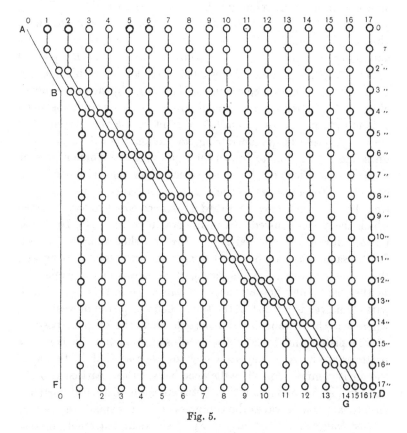

Fig. 5.

intervals τ of time. Thus distances measured vertically downwards from the top row are proportional to the time elapsed since the piston began to move, and distances measured horizontally from left to right from the vertical line through A are proportional to the distances from the original position of the piston. The wave produced by the motion of the piston travels with a velocity v so

that the air particles at a greater distance than vt from A have
not yet been disturbed.   Let the original distances between the
particles shown be each equal to d and suppose that in the time τ
the sound wave travels a distance d so that $d = v\tau$.   Then all the
particles represented by circles which are above the line AD are in
their original positions.   During the interval 0 to 3τ the piston
moves with uniform velocity v' from left to right and its motion is
represented by the line AB.   It then remains at rest as shown by
the vertical line BF.   The air in contact with the piston therefore
comes to rest at B at the time 3τ, and if we draw a line BG parallel
to AD all the air particles represented by circles below this line will
be at rest for the motion communicated to the air by the piston moves
along with velocity v.   The particles represented by circles between
AD and BG are moving from left to right with a velocity equal to
the velocity of the piston during its motion from A to B.   The
lines like 11, 22, 33, etc. drawn through the circles which represent
successive positions of the same particle show the motion of each
particle.   For example, particle No. 10 remains at rest till the
time 10τ and then moves forward a distance equal to the distance
which the piston moved in the interval 0 to 3τ.   No. 10 moves
forward during the interval 10τ to 13τ and then remains at rest.
In the same way No. 3 moves during the interval 3τ to 6τ and
then remains at rest.   Thus each particle moves in the same way
as the piston but not at the same time.   The time when each
particle moves is later than the time when the piston moves by x/v,
where x is the distance of the particle from the original position
of the piston and v the velocity of sound in air.   Where the
particles are moving they are closer together so that in the wave
the air is compressed.   The distance between the particles in the
wave in Fig. 5 is about half that between the undisturbed particles.
In ordinary sound waves the compression is very small, but in very
intense waves it may be large.   Fig. 5 shows an extremely intense
wave in which the motion of the air is large so that it can be clearly
seen on the diagram.

Fig. 6 is a similar diagram showing the motion of the air
particles when the piston moves from A to B during the interval
0 to 3τ and then moves back to its original position during the
interval 3τ to 6τ.

In this case each particle moves from left to right a distance equal to $A'B = 3v'\tau$ and then back to its original position just as the piston does, but at a time x/v later. The backward motion of the piston produces a rarefaction of the air which follows the compression and lies between the lines BG and CE. In the compression

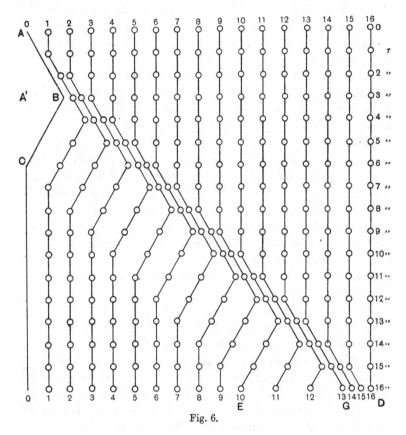

Fig. 6.

the air particles are moving forwards and in the rarefaction backwards with an equal velocity.

The compression is produced by the forward motion because each particle begins to move sooner than the one next after it. For example, No. 9 begins to move τ seconds before No. 10 and in this time moves a distance $v'\tau = \dfrac{A'B}{3}$, where v' is the velocity of the piston, so that its distance from No. 10 is diminished from $d = v\tau$ to $\tau(v - v')$. The rarefaction is produced in the same way by the backward motion. The distance between the particles in the rarefaction is $\tau(v + v')$.

The diagrams shown describe the motion of the air particles produced by the motion of the piston but they do not explain why the air moves in the way described. Let us consider a particular particle, No. 9 say, and show that it is acted on by the forces required to give it the motion described. Up to the time 9τ No. 9 remains at rest and the pressure is the same on both sides of it so that there is no force tending to move it. At the instant 9τ the front of the compression arrives at No. 9, so that just then the pressure is greater behind it than in front; it therefore receives an impulse and starts moving forwards. While it is in the compression, that is, from 9τ to 12τ, the pressure is the same on both sides of it, and so its velocity remains constant. When the back of the compression reaches it the rarefaction is behind it and the compression in front, so that there is a backward impulse on it which converts its forward velocity into a backward velocity. From 12τ to 15τ it is in the rarefaction and moves with uniform velocity backwards, and when the back of the rarefaction reaches it it gets a forward impulse which just stops it, and it then remains at rest in its original position.

In Fig. 7 let there be a wave of compression between the planes AB and CD, and let it be moving from left to right with velocity v. Let the pressure in the wave be p', and let that in the surrounding undisturbed air be p. At the plane CD there is an unbalanced force per unit area equal to $p' - p$. But the plane CD is advancing with velocity v so that if the velocity of the air in the wave is v' the amount of momentum given to the air per second per unit area of CD is equal to $\rho v v'$, where ρ is the density of the undisturbed air.

Hence $$p' - p = \rho v v',$$

since force is equal to rate of change of momentum. In a short time t the plane CD moves forward a distance vt, and the air at CD at the beginning of the time t moves forward a distance $v't$, so that a volume of air vt outside the wave is compressed into a volume $(v-v')\,t$ in the wave. The change of volume per unit volume of the air is therefore equal to

$$\frac{(v-v')-v}{v} = -\frac{v'}{v}.$$

If E denotes the volume elasticity of the air, we have, therefore,

$$p'-p = -E\left(-\frac{v'}{v}\right) = E\frac{v'}{v}.$$

Hence　　　　　　$\rho v v' = E\dfrac{v'}{v}$, or $v = \sqrt{\dfrac{E}{\rho}}.$

It appears therefore that the velocity with which longitudinal sound waves advance through air or any other fluid is equal to the square root of the quotient of the bulk modulus of elasticity by the density of the fluid.

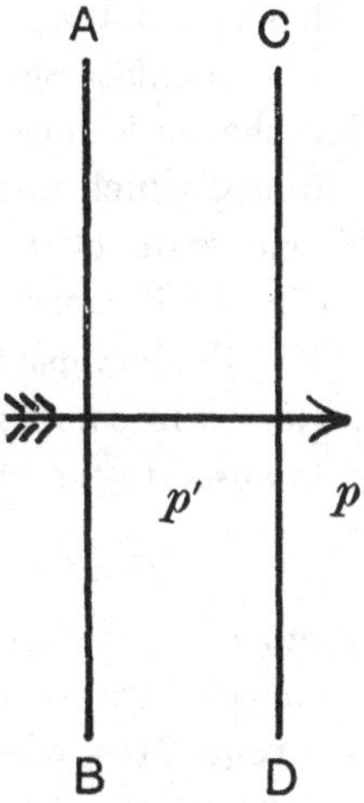

When sound waves pass through a gas the changes of pressure take place so quickly that there is no time for any heat to enter or leave any portion of the gas. If the gas is compressed when a wave passes over it, it therefore gets hotter, which makes its pressure rise more than if its temperature had remained constant. The bulk modulus of elasticity is defined by the equation

$$p'-p = -E\,\frac{V'-V}{V},$$

Fig. 7.

where V' is the volume at pressure p' and V the volume at pressure p, and p' is supposed to be only very slightly greater than p. If the temperature of the gas is supposed constant we have for a gas that obeys Boyle's law

$$p'V' = pV$$

or　　　　　　$(p'-p)\,V = p'\,(V-V').$

Hence　　　　　　$E = -\dfrac{V(p'-p)}{V'-V} = p'.$

If p and p' are nearly equal, which is the case in sound waves of ordinary intensity, we may put $E=p$. The bulk modulus of elasticity of a gas at constant temperature is therefore equal to its pressure. This bulk modulus is called the isothermal elasticity of the gas.

Newton, who first obtained the formula $v = \sqrt{\dfrac{E}{\rho}}$ in the year 1726, tried to calculate the velocity of sound in air by putting $E = p$. The result he obtained did not agree with the observed velocity. Laplace and Poisson in 1807 pointed out that the temperature of the air should not be supposed to be constant, because when the gas is compressed in a sound wave its temperature must rise. Instead of the isothermal elasticity the elasticity when no heat enters or leaves the gas should be used. This is called the adiabatic elasticity of the gas.

To calculate the adiabatic elasticity of a gas we may suppose that the work done on the gas when it is compressed is converted into heat which raises its temperature. Let V denote the volume of one gram of the gas at pressure p. If it is compressed to a volume V' very slightly less than V the work done on it is $p(V - V')$, because for a small change in V the change in p can be neglected in comparison with p. Let the rise of temperature be from absolute temperature θ to θ' so that

$$p(V - V') = JC_v(\theta' - \theta),$$

where C_v is the specific heat of the gas at constant volume and J the mechanical equivalent of heat. We have also $J(C_p - C_v) = R$ (see page 215), where R is the gas constant for one gram of the gas and $pV = R\theta$. Also $p'V' = R\theta'$ where p' is the pressure when the volume is V'. We suppose p and p' are nearly equal. Let $V' - V = c$ and $p' - p = d$, so that

$$p'V' = (p + d)(V + c) = R\theta'$$

or $$pV + pc + Vd + cd = R\theta'.$$

The product cd can be neglected because both c and d are very small, so that, since $pV = R\theta$, we get

$$pc + Vd = R(\theta' - \theta) = J(C_p - C_v)(\theta' - \theta).$$

We have $$pc = JC_v(\theta - \theta'),$$

so that $$Vd = JC_p(\theta' - \theta).$$

These two equations give

$$\frac{Vd}{c} = \frac{pJC_p(\theta' - \theta)}{JC_v(\theta - \theta')} = -p\frac{C_p}{C_v}.$$

The elasticity E is given by the equation

$$d = -E\frac{c}{V} \quad \text{or} \quad E = -\frac{Vd}{c},$$

so that finally

$$E = p\frac{C_p}{C_v}.$$

If $\frac{C_p}{C_v} = \gamma$, then the adiabatic elasticity is equal to γp. For air at $0°$ C. and 76 cms. mercury pressure we have

$$\gamma = 1.41, \quad p = 1.014 \times 10^6 \text{ dynes per sq. cm.}$$

and

$$\rho = 0.001293 \frac{\text{grams}}{\text{c.c.}}.$$

Hence

$$v = \sqrt{\frac{\gamma p}{\rho}} = \sqrt{\frac{1.41 \times 1.014 \times 10^6}{0.001293}},$$

or

$$v = 33260 \frac{\text{cms.}}{\text{sec.}}.$$

This agrees well with the observed velocity 33200, which shows that it is correct to use the adiabatic elasticity γp and not the isothermal elasticity. The gas equation $p_t V_t = p_0 V_0 (1 + \alpha t)$, if V_t is taken to be the volume of one gram, gives

$$\frac{p_t}{\rho_t} = \frac{p_0}{\rho_0}(1 + \alpha t),$$

where ρ_t denotes the density at $t°$ C. Hence

$$v = \sqrt{\frac{\gamma p}{\rho}} = \sqrt{\frac{\gamma p_0}{\rho_0}(1 + \alpha t)}.$$

According to this the velocity of sound should be independent of the pressure and proportional to $\sqrt{1 + \alpha t}$. If t is small this is nearly equal to $1 + \frac{\alpha t}{2}$ or $1 + 0.00184t$. Hence

$$v = 33200(1 + 0.00184t),$$

or

$$v = 33200 + 61t \frac{\text{cms.}}{\text{sec.}}.$$

This formula agrees well with the observed velocities in air. The velocity of sound in other gases can be calculated by the formula $v = \sqrt{\dfrac{\gamma p}{\rho}}$. The equation $v = \sqrt{\dfrac{E}{\rho}}$ also gives the velocity of sound in liquids. For water we have $\rho = 1$ and $E = 2 \times 10^{10}$, hence

$$v = \sqrt{2 \times 10^{10}} = 140000 \,\frac{\text{cms.}}{\text{sec.}},$$

which agrees fairly well with the value found.

CHAPTER III

WAVE TRAINS

WE have seen that when a sound wave is produced in a long cylinder by suddenly moving a piston at one end of the cylinder each particle of air moves in the same way as the piston, but at a time t later than the piston given by $x/v = t$, where x is the distance of the particle from the piston and v is the velocity of the sound. Suppose now that the piston is made to move with a simple harmonic motion of amplitude A and period T. In Fig. 8

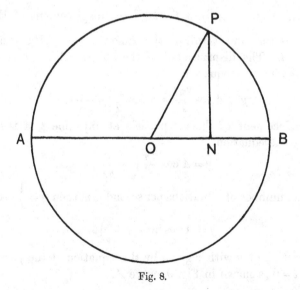

Fig. 8.

let PBA be a circle of radius A, and let P move round the circle with a uniform velocity such that it goes once round in the time T. Let AB be a fixed diameter, and draw PN perpendicular to AB. Then N describes a simple harmonic motion of amplitude A

and period T. Let the angle POB be denoted by ϕ, so that $\phi = 2\pi t/T$, where t is the time measured from the instant when P was at B. For P goes once round in T seconds, so that the angular velocity of OP is $2\pi/T$. We may take the motion of the piston to be the same as the motion of N, so that ON is equal to the displacement of the piston from the middle point of its oscillations. Now $ON = OP \cos \phi = A \cos \dfrac{2\pi t}{T}$. Let $ON = y$, so that

$$y = A \cos \frac{2\pi t}{T}.$$

This equation gives the distance y of the piston from its mean position at any time t when it is moving with a simple harmonic motion of amplitude A and period T. Now consider a particle of air in the cylinder at a distance x from the mean position of the piston. It will move in the same way as the piston, but at a time x/v later. Consequently if $t' = t + \dfrac{x}{v}$ the displacement of the air particle at the time t' will be the same as that of the piston at the time t. The displacement of the air particle at t' is therefore given by the equation

$$y = A \cos \frac{2\pi t}{T} = A \cos \frac{2\pi}{T}\left(t' - \frac{x}{v}\right).$$

The displacement of the air particle at any time t is therefore given by the equation

$$y = A \cos \frac{2\pi}{T}\left(t - \frac{x}{v}\right).$$

If the number of vibrations per second is n, then $n = \dfrac{1}{T}$, so that

$$y = A \cos 2\pi n \left(t - \frac{x}{v}\right).$$

The variation of y with x given by this equation at the particular instant $t = 0$ is shown in Fig. 9, curve A.

When $x = 0$, v/n, $2v/n$, $3v/n$, $4v/n$, etc., we get $y = A$.
When $x = v/2n$, $3v/2n$, $5v/2n$, etc.,   we get $y = -A$.
When $x = v/4n$, $3v/4n$, $5v/4n$, etc.,   we get $y = 0$.

The curve showing the relation between y and x is what is called a cosine curve.

If we take $t = 1/4n$, we get

$$y = A \cos 2\pi n \left(\frac{1}{4n} - \frac{x}{v} \right),$$

or

$$y = A \cos \left(\frac{\pi}{2} - 2\pi n \frac{x}{v} \right).$$

The curve given by this equation is also shown in Fig. 9 and is marked $(4n)^{-1}$. This curve is like the first one, but it is moved

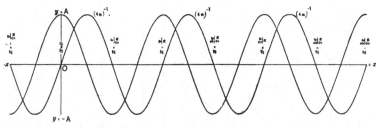

Fig. 9.

from left to right relatively to the first curve a distance $\frac{1}{4} \frac{v}{n}$. In the time $t = \frac{1}{4n}$ the sound waves move a distance s along the cylinder given by $s = tv = v/4n$. Thus we see that the equation

$$y = A \cos 2\pi n \left(t - \frac{x}{v} \right)$$

can be represented by a cosine curve like the curve A moving along from left to right with velocity v. Each particle of air moves like the piston but at a time t later such that $t = x/v$, so the motion travels along with the velocity v. Thus we get a series of waves passing along the cylinder. Each particle of air describes a simple harmonic motion of period T and amplitude A, and the longitudinal displacement y of the air at any time t and any distance x from the piston is given by

$$y = A \cos 2\pi n \left(t - \frac{x}{v} \right).$$

Fig. 10 shows the positions of the air particles at different times represented in the same way as in the previous chapter. The piston is supposed to start moving when $t = 0$, and its motion is represented by the curve 00 from $t = 0$ to $t = 16\tau$. The motions of the particles are represented by the curves 11, 22, 33, etc. We see that a series of compressions with rarefactions between them start from the piston and move along the cylinder. The period of the

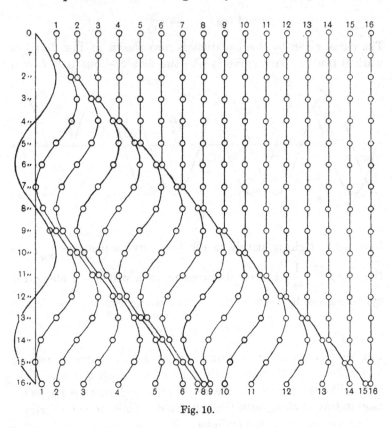

Fig. 10.

oscillation is 8τ and two complete oscillations of the piston are shown. At the time 16τ two complete waves have been formed.

In the curve represented by the equation

$$y = A \cos 2\pi n \left(t - \frac{x}{v} \right)$$

the values of y are repeated when the angle $2\pi n \left(t - \frac{x}{v} \right)$ is increased by 2π. If t is constant, then y is repeated when x is increased by v/n. The distance v/n is called the wave length of the sound. If λ denotes this wave length, then $\lambda = v/n$ or $v = n\lambda$. In one second the piston makes n complete vibrations and therefore gives out n complete waves which extend a distance v, so that $v = n\lambda$. The equation

$$y = A \cos 2\pi n \left(t - \frac{x}{v} \right)$$

may therefore be replaced by

$$y = A \cos \frac{2\pi}{\lambda} (vt - x)$$

or by

$$y = A \cos \frac{2\pi}{\lambda} (x - vt),$$

for changing $vt - x$ to $x - vt$ does not change the displacement y. The equation

$$y = A \cos \frac{2\pi}{\lambda} (x - vt)$$

shows clearly that the waves advance with velocity v, for if x is increased by d to $x + d$, and t at the same time increased from t to $t + \dfrac{d}{v}$, the value of y remains unchanged. A series of waves like that coming from the piston is called a train of waves.

If instead of a piston vibrating at the end of a cylinder we consider a body vibrating in the open air, such as a tuning fork, then a series of sound waves spreads out from it in all directions with the velocity v. The waves are spherical, and as the radius of a wave gets bigger the amplitude in it gets less. A train of spherical waves radiating outwards in all directions from a point can be represented by the equation

$$y = \frac{A}{r} \cos \frac{2\pi}{\lambda} (r - vt),$$

where the amplitude at a distance r from the source is equal to A/r and y is the displacement of an air particle in the direction of the radius r.

The maximum velocity of a particle describing a simple harmonic motion is equal to $2\pi An$, so that its energy is proportional to A^2n^2 or to A^2/λ^2 since $v = n\lambda$. Thus we see that the energy per c.c. in a train of waves is proportional to A^2/λ^2. In the case of spherical waves the energy which starts from the source is spread over a sphere of surface equal to $4\pi r^2$, so that the

energy per c.c. in the waves must fall off inversely as the square
of the distance from the source. The amplitude therefore falls
off inversely as the distance from the source. This is assuming
that no energy is absorbed by the medium as the waves pass
through it. If some is absorbed the amplitude falls off more
rapidly than inversely as the distance.

The propagation of a train of longitudinal waves can be
shown with the model described on page 227. If the first sphere
is moved backwards and forwards with a simple harmonic motion
the compressions and rarefactions can be seen to follow each other
along the row of spheres, and each sphere can be seen to move
backwards and forwards like the first one.

The equation $y = A \cos \dfrac{2\pi}{\lambda}(x - vt)$ gives the displacement

The Pressure in a of the air at any point in a train of sound waves.
Train of Waves. Let p denote the pressure in the undisturbed air
and p' that in the train of waves. We have

$$p' - p = - E\frac{V' - V}{V},$$

where V is the volume of any quantity of the air when undisturbed and V' the volume of the same quantity at the point in
the wave train where the pressure is p'. Consider two planes at
distances x and x' from the origin. Let the longitudinal displacement of the air at x be y, and at x' let it be y'. We have

$$y = A \cos \frac{2\pi}{\lambda}(x - vt)$$

and

$$y' = A \cos \frac{2\pi}{\lambda}(x' - vt).$$

The volume of the air when undisturbed between the two planes
is equal to $x' - x$ per unit area of the planes. When the air
is disturbed by the train of waves this volume becomes
$(x' + y') - (x + y)$, because the air which was at x is displaced to
$x + y$, and the air which was at x' is displaced to $x' + y'$. If the
two planes are taken very near together we may put $V = x' - x$
and $V' = (x' + y') - (x + y)$, so that $V' - V = y' - y$.

Hence

$$p' - p = - E\frac{y' - y}{x' - x}.$$

Now

$$y' - y = A \cos \frac{2\pi}{\lambda} (x' - vt) - A \cos \frac{2\pi}{\lambda} (x - vt),$$

so that since

$$\cos D - \cos C = 2 \sin \frac{C + D}{2} \sin \frac{C - D}{2}$$

we get

$$y' - y = 2A \sin \frac{2\pi}{\lambda} \left(\frac{x' + x}{2} - vt \right) \sin \frac{\pi}{\lambda} (x - x').$$

But $x - x'$ is very small, so that

$$\frac{x' + x}{2} = x$$

and

$$\sin \frac{\pi}{\lambda} (x - x') = \frac{\pi}{\lambda} (x - x').$$

Therefore

$$\frac{y' - y}{x - x'} = 2A \frac{\pi}{\lambda} \sin \frac{2\pi}{\lambda} (x - vt).$$

Hence

$$p' - p = 2EA \frac{\pi}{\lambda} \sin \frac{2\pi}{\lambda} (x - vt)$$

or

$$p' - p = P \sin \frac{2\pi}{\lambda} (x - vt),$$

where

$$P = 2EA \frac{\pi}{\lambda}.$$

It appears therefore that in a train of waves if

$$y = A \cos \frac{2\pi}{\lambda} (x - vt)$$

represents the displacements, then

$$p' - p = P \sin \frac{2\pi}{\lambda} (x - vt)$$

represents the change of pressure.

Fig. 11 shows the variation of the displacement y along the train of waves and also the variation of the pressure at the particular instant when $t = 0$. The maximum values of the pressure occur a quarter of a wave length ahead of the maximum values of the displacement.

We have

$$\sin \frac{2\pi}{\lambda} (x - vt) = \cos \left\{ \frac{2\pi}{\lambda} (x - vt) - \frac{\pi}{2} \right\},$$

so that
$$y = A \cos \frac{2\pi}{\lambda} (x - vt)$$

and
$$p' - p = P \cos \left\{ \frac{2\pi}{\lambda} (x - vt) - \frac{\pi}{2} \right\}.$$

The phase of the pressure variation is therefore 90° or $\frac{\pi}{2}$ ahead of
that of the displacement. If the pressure curve were moved back
$\lambda/4$ and P made equal to A the two curves would coincide. We
see from Fig. 11 that the pressure is a maximum where the

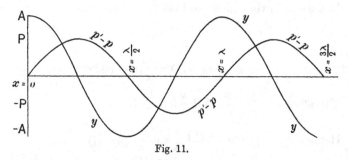

Fig. 11.

displacement in the forward direction is increasing most rapidly
and a minimum where it is diminishing most rapidly, for as
the displacement curve moves along from left to right the displacement at a fixed point rises where it slopes downwards and
falls where it slopes upwards. The same thing may be seen
by studying Fig. 10 (page 240).

CHAPTER IV

MUSICAL NOTES

THE sound produced by a body vibrating with a simple harmonic motion is a musical note. What is called the pitch depends on the number of vibrations per second, or the frequency. When the frequency is great the pitch is high, and when the frequency is small the pitch of the note is low. It is found that the sound produced by a vibrating body like a tuning fork is not audible unless the frequency lies between certain limits. These limits are different for different people. For most people they are about 30 to 30,000 vibrations per second. A tuning fork or other vibrating body vibrating more than 30,000 times per second produces no sound audible by most people and a fork making less than 30 vibrations per second is also inaudible. Vibrating bodies with frequencies outside the limits of audibility can be shown to produce trains of waves in the air; but the trains do not affect the ear.

The motion of the air at any point produced by a body like a tuning fork moving in a simple harmonic motion is also a simple harmonic motion of the same frequency as the fork. A simple harmonic motion is completely determined when we know its amplitude, frequency and phase. If the motion is represented by the equation

$$y = A \cos (2\pi nt + \alpha),$$

so that, when $t = 0$, $y = A \cos \alpha$, then the angle α is called the phase of the vibration. The phase is usually of no importance when we are dealing with only one vibration, but when the result of adding two or more vibrations together has to be considered the phases may be important. When dealing with only one vibration we can always reckon the time from an instant when $y = A$, so that

$\cos \alpha = 1$ and the phase is zero. The loudness or intensity of a musical note depends on the amplitude of the vibration, and the pitch on the frequency. The relative loudness of notes of different frequencies is difficult to estimate and we do not know the relative amplitudes of notes of different pitch which seem equally loud. All we can say is that if a note seems louder than another of equal frequency then it has the greater amplitude.

The energy per cubic cm. in the air due to a train of waves is proportional to $A^2 n^2$, where A is the amplitude and n the frequency.

If a series of tuning forks having frequencies proportional to the numbers 1, 2, 3, 4, 5, 6, etc. are sounded together, then it is found that the sound produced seems to have the pitch of the fork of lowest frequency, but the quality of the sound depends on the relative intensities of the sounds due to the different forks. The sounds of the different forks blend together and seem to consist of only one musical note. For example, if forks with frequencies 128, 256, 384 and 512 are sounded together, the sound produced would be considered by a musician to have the same pitch as the sound produced by the 128 fork sounding by itself. It is found that the sounds produced by most musical instruments consist of the sum of a series of vibrations with frequencies proportional to 1, 2, 3, 4, etc. The vibration of lowest frequency is called the fundamental tone and determines the pitch. The other vibrations are called the harmonics of the fundamental tone. The vibration with frequency equal to twice that of the fundamental is called the first harmonic, that with three times the second harmonic, and so on. The quality of the sounds emitted by musical instruments depends on the relative intensities of the harmonics present. A musical note like that produced by a tuning fork, which consists of the fundamental alone, is sometimes called a pure tone or a simple tone. The loudness or intensity, pitch, and quality of a musical note are therefore determined respectively by the amplitude, frequency of the fundamental, and relative intensities of the fundamental and harmonics. It is found that the relative phases of the fundamental and harmonics make no difference to the quality of the sound. In later chapters we shall consider the notes emitted by various musical instruments, and the harmonics which they contain.

The frequency of a musical note can be found approximately
with an instrument called a siren. A simple
Determination of Frequency. form of a siren is shown in Fig. 12. This
consists of a metal disk *D*, about 20 cms. in diameter, mounted
on a horizontal shaft, which can be made to rotate by means
of a small electric motor *M*. The number of revolutions made
by the disk is shown by a revolution-counter driven by the
shaft. The speed of the disk can be varied by changing
the current through the armature or magnets of the motor.
A circular row of equidistant holes is bored in the disk near

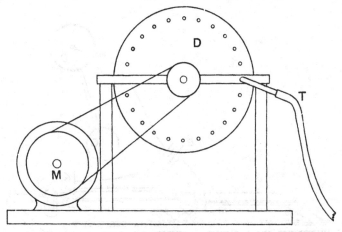

Fig. 12.

the circumference, as shown. A tube is supported at right
angles to the disk and close to it, so that the holes in the disk
move past the end of the tube. Air can be blown through the
tube by means of a rubber tube *T*. Suppose that there are *n* holes
in the disk and that it makes *m* revolutions per second. Then *nm*
holes pass in front of the tube per second. If air is blown into
the tube a puff of air escapes through each hole so that there are
nm puffs per second. Each puff produces a wave of compression
followed by a rarefaction so that we get *nm* complete vibrations
produced in the air per second. The siren therefore produces
a musical note of frequency *nm*. To find the frequency of any
note with a siren, the speed of the disk is adjusted until the note

given by the siren has the same pitch. The speed is then kept
constant, and the number N of revolutions in a known time t is
found. The frequency is then equal to $\dfrac{nN}{t}$.

The frequency of a tuning fork can be found with the apparatus
shown in Fig. 13. The tuning fork F is supported by a pillar H
fixed to a wooden base. A glass plate PP can slide along the
base between two guides. A light metal pointer S is fastened to
one prong of the fork and just touches the glass plate. The plate
is coated with lamp black by holding it in a smoky flame, and if it
is made to slide along the base while the fork is vibrating the

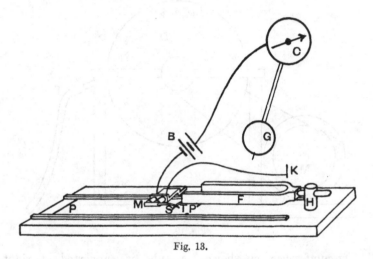

Fig. 13.

pointer S traces a wavy line in the lamp black. Another light
pointer T also just touches the plate close to the pointer S. The
pointer T has a small piece of iron attached to it which is attracted
by a small electromagnet M when a current is passed through the
magnet. One of the wires from the magnet leads to the top of
the pendulum G of the clock C and the other wire is connected to
a battery B and a spring K which the lower end of the pendulum
just touches when at its lowest point. The clock pendulum has a
period of two seconds, so that it touches K once every second. This
makes the pointer T move suddenly sideways once every second.
When the glass plate is pulled along between its guides the

vibrating tuning fork makes a wavy line, and close beside this the pointer T makes a straight line with kinks in it. The time from one kink to the next is one second. It is easy to count the number of waves in the wavy line made by S between two kinks made by T, which is equal to the number of vibrations made by the fork in one second. By doing this several times and taking the mean of the results the frequency of the fork can be found accurately.

In experiments on sound it is often convenient to be able to keep a tuning fork steadily vibrating. This can be done by means of an electrical device shown in Fig. 14. The fork F is supported by a pillar R fixed to a wooden base B. A short electromagnet M is fixed between the prongs of the fork but does not touch them.

Electrically Maintained Tuning Forks.

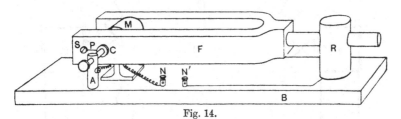

Fig. 14.

One prong has a platinum wire P fixed to it by a small screw S. This wire projects slightly from the fork and just touches the face of the disk C carried by a screw which passes through a pillar A fixed to the base. The face of the disk C is made of platinum. One of the wires from the magnet leads to a binding screw N and the other to the pillar A. A wire leads from the pillar R to the other binding screw N'. If N' and N are connected to two or three dry cells the current passes from N' to R and hence to S and P. From P it flows to C and then through the magnet to N. The magnet draws the prongs of the fork together and so the wire P breaks its contact with the disk C. This stops the current, so that the magnet stops attracting the prongs of the fork, which move back so that P and C again make contact. In this way the fork is kept vibrating steadily. The electric current in the wires leading to N' and N is an intermittent current consisting of as many flows of electricity per second as the fork makes complete vibrations.

The intermittent current can be used to drive a simple kind of electric motor called a phonic wheel, by means of which the frequency of the fork can be very accurately found.

Fig. 15 shows a phonic wheel. It consists of an iron wheel C mounted on a shaft carried by bearings. The wheel has a number of equidistant projections or cogs on it. Two small electromagnets A and B are supported at opposite ends of a diameter of the wheel so that the cogs almost touch the magnets when the wheel rotates. If the intermittent current from an electrically maintained fork and battery are passed through the magnets of the phonic wheel

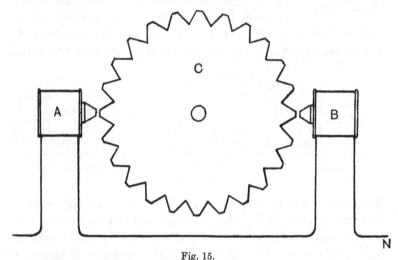

Fig. 15.

and the wheel started at the proper speed, the magnets keep the wheel running. Suppose the fork makes 200 vibrations per second, then the magnets of the phonic wheel will be excited 200 times per second. If the wheel is started so that 200 cogs pass each magnet per second, then as each cog is coming to the magnet it will be attracted for a moment, and so the wheel will be kept going. The shaft of the wheel drives a revolution-counter, so that the number N of revolutions it makes in a known time t can be found. If the wheel has m cogs, then the frequency of the fork is $\dfrac{mN}{t}$.

The wheel can be kept going for a long time, say one hour,

which can be easily found to within one second or one part in
3600. It is easy with a phonic wheel to get the frequency of
a fork to within one part in 3000. The revolution-counter is
sometimes arranged so that it rings a bell at every 100 revolutions. The times when the bell rings are then observed on an
accurate clock. Suppose the time at which the bell rings is
observed for say 20 successive rings. Let the observed times
be denoted by $t_1, t_2, t_3 \ldots t_{20}$. A good way to calculate the frequency from these times is to find the mean of $t_{11} - t_1$, $t_{12} - t_2$,
$t_{13} - t_3$, $t_{14} - t_4$, $t_{15} - t_5$, $t_{16} - t_6$, $t_{17} - t_7$, $t_{18} - t_8$, $t_{19} - t_9$ and $t_{20} - t_{10}$.

The mean of these intervals is taken to be the time during which
the fork makes $1000m$ complete vibrations. For example, with
a wheel having 25 cogs if the observed times were 0 sec., 10·5 secs.,
20·2 secs., 29·8 secs., 41 secs., 50·3 secs., 59·5 secs., 70 secs., 80·1 secs.,
89·9 secs., 100 secs., 110 secs., 121 secs., 130·2 secs., 140 secs.,
149·5 secs., 160 secs., 170·5 secs., 180 secs. and 190·5 secs., we
should have

$$
\begin{aligned}
100 \ - 0 \ &= 100\cdot0 \\
110 \ -10\cdot5 &= 99\cdot5 \\
121 \ -20\cdot2 &= 100\cdot8 \\
130\cdot2 - 29\cdot8 &= 100\cdot4 \\
140 \ -41 \ &= 99\cdot0 \\
149\cdot5 - 50\cdot3 &= 99\cdot2 \\
160 \ -59\cdot5 &= 100\cdot5 \\
170\cdot5 - 70 \ &= 100\cdot5 \\
180 \ -80\cdot1 &= 99\cdot9 \\
190\cdot5 - 89\cdot9 &= 100\cdot6
\end{aligned}
$$

Sum 1000·4   Mean 100·04 secs.

$$\text{Frequency} = \frac{2500 \times 10}{100\cdot04} = 249\cdot90 \text{ vibrations per sec.}$$

It is found that the frequency of a steel tuning fork remains
very constant if the fork is carefully preserved. The frequency
varies slightly with the temperature of the fork.

If two tuning forks having frequencies proportional to small
whole numbers such as 1 and 2 or 2 and 3 are
sounded together it is found that the two notes
blend together in a way pleasing to musicians. If the two forks
are sounded in succession a certain relation between the pitches
can be recognised. The notes used in music should have such

The Musical Scale.

simple relations between their frequencies. The ratio of the
frequencies is called the *interval* between the notes. The interval
2/1 is called an octave. For example a note of frequency 500 is
called the octave of a note of frequency 250. The following table
gives the names and frequency ratios of other intervals used in
music:

Interval			Frequency Ratio
Octave ...	...	...	2 : 1
Fifth ...	...	...	3 : 2
Fourth ...	...	...	4 : 3
Major third	...	...	5 : 4
Minor third	...	...	6 : 5
Major sixth	...	...	5 : 3
Minor sixth	...	...	8 : 5

The musical scale of notes called the diatonic scale is obtained
by starting with a definite note called the key note and choosing a
series of notes between which and the key note the above ratios
hold.

Three notes with frequencies proportional to 4, 5 and 6 form
what is called a major chord. A minor chord consists of three
notes with frequencies proportional to 10, 12 and 15.

If the key note of the diatonic scale has a frequency n, then
its octave is $2n$. Between n and $2n$ in the diatonic scale there
are six notes with frequencies $\frac{9}{8}n, \frac{5}{4}n, \frac{4}{3}n, \frac{3}{2}n, \frac{5}{3}n$, and $\frac{15}{8}n$. The
next octave higher from $2n$ to $4n$ contains six notes between $2n$
and $4n$ with frequencies double those between n and $2n$, and so on
for the other octaves. The frequency of the key note is usually
taken to be 256 for scientific purposes. In musical circles it
is generally taken to be somewhat higher. The diatonic scale is
only used with instruments like the violin which can give notes of
any frequency. The scale used on the piano contains eleven notes
between n and $2n$ with frequencies adjusted so that the ratio of
any note to the next is the same in all cases. The value of this
ratio is therefore $2^{\frac{1}{12}}$. The ratios of the frequencies on this scale
are not exactly those of small whole numbers, and the music of
the piano is less pleasing to musicians than music based on the
diatonic scale.

In scientific work it is best to specify musical notes by stating their frequencies and not by giving the letters used to designate them by musicians. The middle C on the piano usually has a frequency rather greater than 256.

REFERENCES

Sound and Music, Sedley Taylor.
Sound, Poynting and J. J. Thomson.

CHAPTER V

REFLEXION, REFRACTION, INTERFERENCE OF SOUND AND
COMPOSITION OF PERPENDICULAR VIBRATIONS

When a sound wave in air comes to the surface of a solid body like a wall it is reflected from the surface. Fig. 16 represents a photograph of a spherical sound wave produced by an electric spark being reflected by a plane surface. The spark occurred at S and WW is the spherical wave spreading out with S as centre.

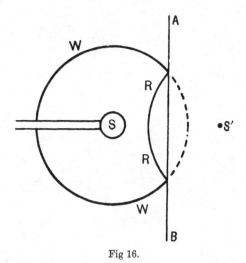

Fig 16.

AB is a plane metal plate and RR is the part of the wave which has been reflected. RR is part of a sphere with its centre at S'. The line SS' is perpendicular to AB and the distance from AB to S' is equal to the distance from AB to S. The dotted line shows where the wave would have been if the plane had not been there.

If a plane sound wave meets a plane surface parallel to the wave its velocity is reversed and it moves back along the path by which it came to the surface.

Echoes are due to the reflexion of sound. If a gun is fired in front of a steep cliff or mountain the sound wave produced by the gun may be reflected back to the gun so that anyone near the gun hears the sound of the gun repeated after a short interval. Whenever an echo is observed it is found that some surface from which the sound is reflected back is present. Sound is reflected best from smooth rigid surfaces. Soft substances like felt do not reflect sound much. The echoes in large rooms due to reflexions from the walls can be stopped to a great extent by covering the walls with curtains made of felt or other soft thick cloth. Large smooth surfaces should be avoided in rooms where no echo is desirable. Rooms intended for public speaking or for concerts should be designed so as to prevent echoes in them as much as possible. The reflexion of sound can be studied experimentally on a small scale by using what is called a sensitive flame to detect the sound. A sensitive flame is a coal gas flame burning at a jet consisting simply of a round hole about two millimetres in diameter at the end of a tube. The hole points vertically upwards and the pressure of the gas supply is increased until the long thin pointed flame is just on the point of becoming unstable and roaring. If a note of very high pitch is sounded the flame bobs down to about half its previous height and roars. To get sufficient gas pressure it is necessary to use gas from a gas bag loaded with heavy weights.

As a source of sound a very small whistle may be used. If the whistle is put in front of a concave mirror the sound can be reflected from the mirror in a parallel beam which can be focused on the sensitive flame by another mirror as shown in Fig. 17. W is the whistle which can be blown by a tube T. M is a concave mirror. The direction of motion of the sound waves from the whistle is shown by the lines marked with arrow heads. The sound is concentrated on the flame FF close to the burner by means of a second mirror N. B is the gas burner and G a rubber tube leading to the gas bag. The mirrors should be about 18 inches in diameter and may be placed 20 feet or more apart. If a large plane sheet of metal is put up between the two mirrors it can be shown that it stops the sound from getting to the second mirror. If the sheet of metal is placed in a vertical plane inclined to the direction of the beam of sound the beam is reflected from it and can be detected by means of the flame and second mirror if these are moved round the sheet until the place where the flame is strongly affected is found.

Thus if the sheet is placed so that it makes an angle of 45° with the beam from the first mirror the reflected beam will be

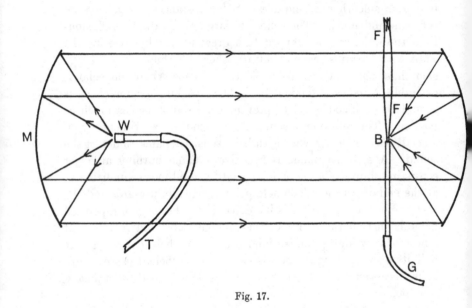

Fig. 17.

found along a direction at right angles to that of the beam before reflexion.

The beam before reflexion may be called the incident beam of sound. It is found that the incident and reflected beams are equally inclined to the reflecting surface.

The velocity of sound in a heavy gas like carbon dioxide is less than in air; in a light gas like hydrogen it is

Refraction of Sound.

greater. The formula $v = \sqrt{\dfrac{\gamma p}{\rho}}$ gives the following velocities at 0° C.

Gas			Velocity
Air ...	...	...	33240 cms./sec.
Carbon dioxide	...		26000 „
Hydrogen ...	...	...	128600 „

The smaller velocity in carbon dioxide can be shown by means of the apparatus represented in Fig. 18. AB is a metal ring about two feet in diameter to which two very thin sheets of india-rubber are fastened. Carbon dioxide gas is pumped into the space between the rubber sheets, so that they are blown out as shown and form a lens-shaped body. W is the whistle and F the sensitive flame. The spherical sound waves like CD spreading out from W pass through the carbon dioxide lens. In the lens the waves travel more slowly than in air, so that at the middle of the lens where it is thickest the waves are retarded relatively to the parts of

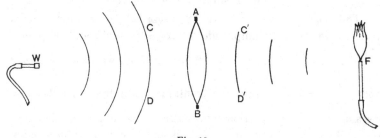

Fig. 18.

the waves which are further from the middle. The result is that when the relative positions of W, AB and F are properly adjusted the convex waves become nearly plane inside the lens, and on coming out on the other side become concave like $C'D'$ and converge to a focus at F. The change of direction of the sound waves in passing through the lens is called refraction. The theory of reflexion and refraction will be more fully discussed in the part of this book dealing with Light. Instead of the lens made with rubber sheets a large soap bubble filled with carbon dioxide can be used.

Interference of Sound.   When two sound waves are passing over a point the displacement of the air at the point is the resultant of the displacements due to the two waves. The change of pressure at the point is the sum of the changes due

to the two waves.   Suppose there are two trains of waves passing
through the air.   Let the pressure variations due to them at a
fixed point be given by the equations

$$p' - p = P \sin \frac{2\pi}{\lambda'} vt$$

and $$p'' - p = P' \sin \frac{2\pi}{\lambda''} vt.$$

The resulting pressure variation will be given by

$$p' - p + p'' - p = P \sin \frac{2\pi}{\lambda'} vt + P' \sin \frac{2\pi}{\lambda''} vt.$$

Let $\frac{v}{\lambda'} = n'$ and $\frac{v}{\lambda''} = n''$, so that n' and n'' are the number of
vibrations per second in the wave trains.   Also let $P' = P$ so that

$$p' - p + p'' - p = P (\sin 2\pi n't + \sin 2\pi n''t)$$
$$= 2P \sin \pi t (n' + n'') \cos \pi t (n' - n'').$$

The last expression may be regarded as representing a
pressure variation of amplitude $2P \cos \pi t (n' - n'')$ and frequency
$\frac{n' + n''}{2}$.   For example if $n' = 202$ and $n'' = 200$ we have $n' - n'' = 2$
and $\frac{n' + n''}{2} = 201$, so that the resultant has a frequency 201 and
its amplitude is proportional to $\cos 2\pi t$.   Its amplitude therefore
is equal to zero twice every second and has a maximum value
positive or negative twice every second.

When two tuning forks of nearly equal frequencies are sounded
together the sound produced rises and falls in intensity.   Each
maximum of intensity is called a beat.   We see from the foregoing
calculation that the number of beats per second is equal to the
difference between the frequencies of the two forks.   The beats
are said to be due to interference between the two sounds.   The
way in which the beats are produced may be explained simply as
follows.   When the pressure changes due to both forks are in the
same direction the resultant pressure changes are greater than
when they are in opposite directions.   If one fork makes n
vibrations per second and the other m, then the first gains a
whole vibration on the other in $\frac{1}{n - m}$ second.   The time between
a maximum of sound and the next minimum is the time in which

one gains half a complete vibration on the other, for the time between the maximum increase of pressure due to one fork and the next maximum decrease is the time of half a vibration. The time between successive maxima of intensity is therefore the time in which one fork gains a complete vibration on the other, so that there are $n - m$ beats per second.

Fig. 19 shows graphically the result of adding together two vibrations of nearly equal frequencies. Curve A shows one vibration and curve B the other. Curve C shows the sum of the vibrations represented by A and B. The time is proportional to the distance measured from left to right. At $t = 0$ the two vibrations agree in phase. The period of A is 4 and its amplitude 2, while the period of B is 6 and its amplitude also 2. At $t = 0$ the amplitude of C is $2 + 2 = 4$. At $t = 6$, A has made $1\frac{1}{2}$ vibrations and B only one, so that A has gained half a vibration and the amplitude of C is zero. At $t = 12$, A has made three vibrations and

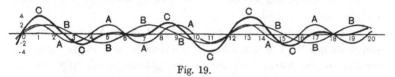

Fig. 19.

B two, so that A has gained a whole vibration on B and the amplitude of C is again 4. The interval between two beats is therefore 12. The number of beats per unit time is $\frac{1}{12} = \frac{1}{4} - \frac{1}{6}$. At $t = 18$, A has made $4\frac{1}{2}$ vibrations and B three, so that A is $1\frac{1}{2}$ vibrations ahead of B and the amplitude of C is again zero.

The difference between the frequencies of two notes of nearly equal frequency can be found by counting the beats when both are sounding. Beats can also be heard when a note and another of nearly double frequency are sounding. The beats are due to interference between the fundamental of the second note and the first harmonic of the first. If one fork has a frequency n and the other m such that m is nearly double n, then the number of beats per second is $2n - m$.

Two tuning forks can be adjusted so that their frequencies are very nearly equal by attaching small weights to the prongs of the one with the higher frequency until the interval between successive beats becomes very long.

If we have a series of forks, each one having a slightly higher frequency than the one before it in the series, and extending over an octave, the frequencies of all the forks can be found by counting the beats between each pair of adjacent forks in the series. For example, suppose there are 33 forks in the series and each one gives 4 beats per second with the one with the next higher frequency; also let the last one give the octave of the first one. Then if the frequency of the first fork is n_1, that of the second n_2, that of the third n_3 and so on, we have the 32 equations:

$$n_2 - n_1 = 4$$
$$n_3 - n_2 = 4$$
$$\vdots \qquad \vdots$$
$$n_{32} - n_{31} = 4$$
$$n_{33} - n_{32} = 4$$

Adding these up we get $n_{33} - n_1 = 128$. But $n_{33} = 2n_1$ so that $n_1 = 128$ and $n_{33} = 256$.

When a train of waves is reflected from a plane surface perpendicular to the direction of motion of the waves, that is, parallel to the waves themselves, we get an interesting case of interference of sound waves. A train of waves travelling along a tube may be reflected from a closed or from an open end of the tube, and then the waves travel back along the tube in the opposite direction and interfere with the waves coming towards the end of the tube.

Suppose we have a train of waves the displacements in which are given by the equation

$$A \cos \frac{2\pi}{\lambda} (x + vt),$$

which represents waves travelling from right to left. If these are reflected so that they go back from left to right, the displacements in the reflected train may be represented by

$$y' = A' \cos \frac{2\pi}{\lambda} (x - vt + c),$$

for the velocity of the waves has been changed from v to $-v$. The constant c is added because the phase of the reflected waves at any point need not be the same as the phase of the waves before reflexion. Let the reflecting surface be at $x = 0$. Then at $x = 0$ the air particles are in contact with the reflecting surface which we shall suppose to be rigid, so that at $x = 0$ the resultant longitudinal displacement of the air must be zero.

The resultant displacement at any point is $y + y'$, so that at $x = 0$ we have

$$0 = y + y' = A \cos \frac{2\pi vt}{\lambda} + A' \cos \frac{2\pi}{\lambda} (c - vt).$$

This equation must be true whatever the value of t, so that it is clear that $A' = -A$ and $c = 0$. We have therefore at any point at which x is greater than 0,

$$y + y' = A \left\{ \cos \frac{2\pi}{\lambda} (x + vt) - \cos \frac{2\pi}{\lambda} (x - vt) \right\}$$

$$= -2A \sin \frac{2\pi}{\lambda} x \sin \frac{2\pi}{\lambda} vt.$$

This shows that at any point there is a simple harmonic vibration of amplitude $2A \sin \frac{2\pi}{\lambda} x$. If $x = \frac{n\lambda}{2}$ where $n = 0, 1, 2$, etc., then $\sin \frac{2\pi}{\lambda} x = \sin n\pi = 0$. Thus at the reflecting surface and at any plane distant from it by a whole number of half wave lengths the air remains at rest. If $x = (2n + 1) \frac{\lambda}{4}$ then $y + y' = 2A \sin \frac{2\pi}{\lambda} vt$ so that then there is a vibration of the maximum amplitude $2A$. The planes where the air remains at rest are called the nodes.

The reflexion of the train of waves reverses the direction of the displacements in it and also the direction of propagation. A displacement in the direction of the propagation remains a displacement in the direction of propagation after the reflexion. At a point half a wave length from the reflecting surface the reflected waves have travelled a whole wave length further than the incident waves, so that if the direction of the displacement had not been reversed the two wave trains would agree in phase and reinforce each other. Owing to the reversal on reflexion the two displacements at half a wave length from the surface are equal and opposite and so always annul each other. The motion of the air particles when a train of waves is reflected by a rigid surface parallel to the waves is shown in Fig. 20. The reflecting surface is at A and a row of 15 equidistant air particles numbered 1 to 15 is shown. The top row shows the positions of the particles when $t=0$ and they are in their undisturbed positions. The following 12 horizontal rows show the same 15 particles at the ends of 12 successive equal intervals of time τ. The time of a complete vibration is equal to 12τ, and the wave length is equal to twelve times the distance d between the successive particles when undisturbed.

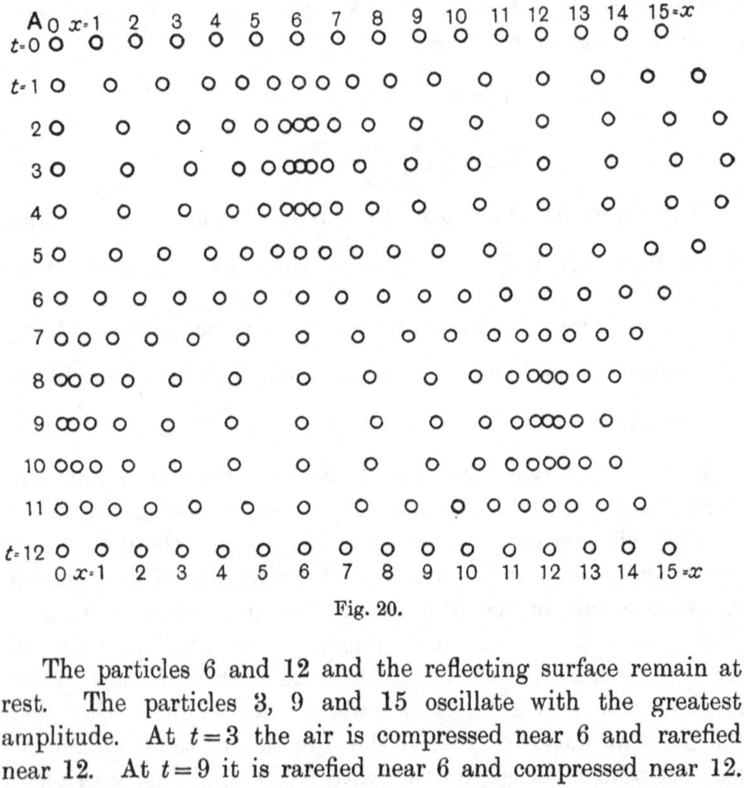

Fig. 20.

The particles 6 and 12 and the reflecting surface remain at rest. The particles 3, 9 and 15 oscillate with the greatest amplitude. At $t=3$ the air is compressed near 6 and rarefied near 12. At $t=9$ it is rarefied near 6 and compressed near 12. We see that the maximum pressure variation occurs at 6, 12 and at the reflecting surface where the air remains at rest. The places of zero pressure variation are 3, 9, and 15 where the motion is greatest. The curves 22, 33, 44 are roughly parallel, showing that little compression or rarefaction occurs between the particles

2 and 4. It can be easily verified that the representation of the motion shown in Fig. 20 agrees with that given by the equation

$$y + y' = -2A \sin \frac{2\pi x}{\lambda} \sin \frac{2\pi vt}{\lambda}.$$

The motion due to two trains of waves of equal amplitudes and frequencies, moving in opposite directions, is called a stationary vibration because of the way in which the motion occurs between fixed planes half a wave length apart.

An apparatus invented by Kundt enables the formation of stationary waves in a tube to be very clearly demonstrated and accurately studied. A simple form of this apparatus is shown in Fig. 21. TT' is a glass tube about 4 cms. in diameter and 100 cms. long. This tube rests on two V-shaped blocks C, C' fastened to a wooden base B. A steel rod RR' about 0·7 cm. in diameter and 100 cms. long is supported at its middle point by a pillar S.

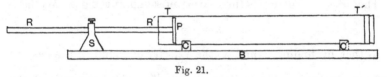

Fig. 21.

One end of the rod projects a few cms. into the glass tube, and carries a light disk of cork or cardboard P. This disk is about 0·5 cm. smaller in diameter than the glass tube and should not touch the glass. The glass tube may be closed by a cork at T'. The distance PT' can be varied by sliding the tube along over the blocks C and C'. A little dry lycopodium powder is spread along the bottom of the glass tube between P and T'. The glass tube should be clean and dry.

The rod RR' can be made to vibrate longitudinally by rubbing it near R with a piece of soft leather coated with powdered rosin. The middle point of the rod remains fixed, and the ends vibrate so that the length of the rod oscillates between small limits. The disk P is thus made to move backwards and forwards parallel to the length of the tube TT' with a simple harmonic motion of high frequency and small amplitude. The motion of the disk produces a train of waves in the air in the tube, which is reflected back from the cork at T' and is then reflected from the disk at P, and so on. If the distance between P and T' is properly adjusted, the waves reflected from P coincide in phase with those produced by the motion of P and the air in the tube is set into a stationary vibration of great intensity. Where the air remains at rest the lycopodium powder is not disturbed, but where the air vibrates the powder is blown about. The result of this is that most of the powder soon collects into little heaps at the points where the air remains at rest. There is a heap close to the cork and one close to the disk, and a series of equidistant heaps between these, half a wave length apart. If the distance between the end heaps is measured and divided by the number of half wave intervals, we get half the wave length of the note emitted by the vibrating disk. We have $v = n\lambda$, where v is the velocity of sound in air, n the frequency, and λ the wave length, so that n can be calculated. If the tube is filled with another gas instead of air, the wave length can be found in the same way. Let it be λ'. Then $v' = n\lambda'$ where v' is the velocity of sound in the gas. We have

$$n = \frac{v}{\lambda} = \frac{v'}{\lambda'},$$

so that by finding λ and λ' the velocity of sound in the gas can be determined. The equation $v = \sqrt{\dfrac{\gamma p}{\rho}}$ enables the ratio of the specific heat at constant pressure to that at constant volume for any gas to be calculated when v is known. Kundt's apparatus is often used to find the value of γ for a gas.

When a train of waves travelling along a tube comes to an open end of the tube, the waves are reflected from the open end and go back into the tube so that very little of the sound escapes. If the cork at T' in the Kundt's apparatus is removed the stationary waves and heaps of powder can still be obtained if the distance between the disk and the end of the tube is adjusted properly. The heap nearest to the open end is found to be only about a quarter of a wave length from it instead of half a wave length as it is when the cork is used. The spaces between the heaps are half wave lengths as before.

Let the pressure change in a train of waves travelling along a pipe from right to left be represented by the equation

$$p' - p = P \cos \frac{2\pi}{\lambda}(x + vt),$$

and suppose the end of the pipe is at $x = 0$ and is open. Let the reflected train of waves going from left to right in the pipe be represented by

$$p'' - p = P' \cos \frac{2\pi}{\lambda} (x - vt + c).$$

At the open end the air is not confined by the walls of the tube, so that it is much freer to move than in the tube. The pressure at the open end will therefore differ very little from the normal pressure p, so that at $x = 0$ we have approximately

$$0 = p' - p + p'' - p = P \cos \frac{2\pi vt}{\lambda} + P' \cos \frac{2\pi}{\lambda} (c - vt).$$

Hence $\qquad\qquad P' = - P$ and $c = 0$.

Hence at any point in the tube

$$p' - p + p'' - p = P \left\{ \cos \frac{2\pi}{\lambda} (x + vt) - \cos \frac{2\pi}{\lambda} (x - vt) \right\}$$

$$= - 2P \sin \frac{2\pi x}{\lambda} \cdot \sin \frac{2\pi vt}{\lambda}.$$

This equation is exactly similar to the equation

$$y + y' = - 2A \sin \frac{2\pi x}{\lambda} \cdot \sin \frac{2\pi vt}{\lambda},$$

which was found to represent the displacements in the closed tube. It appears that the pressure variation is zero at the open end of the tube and at a series of equidistant planes half a wave length apart. Half way between these planes of zero pressure variation there are the planes of zero displacement. There is therefore a plane of zero displacement at a distance $\lambda/4$ from the open end and a series of others at $\frac{5}{4}\lambda$, $\frac{9}{4}\lambda$, $\frac{13}{4}\lambda$, etc. from the open end. Fig. 20 may be taken to represent the motion of the air particles in a tube with an open end if the open end is supposed to be at the particle 15. The planes of zero displacement are at 12, 6 and 0, and the planes of zero pressure variation at 15, 9 and 3.

Stationary vibrations can also be obtained and examined with the apparatus shown in Fig. 22. MM is a large concave reflector which reflects the sound from a high pitched whistle W so as to form a parallel beam of sound. The beam is reflected from a plane metal sheet PP so that it returns along its path. Stationary vibrations are produced between the mirror and plane PP by the incident and reflected waves. The positions of the planes of zero displacement can be found with a sensitive flame *FF*, which is not affected at these planes but roars when at a plane of maximum displacement and zero pressure variation. In this way the wave length of the sound produced by the whistle can be determined.

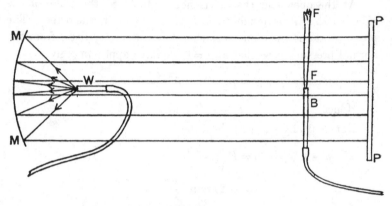

Fig. 22.

If a high pitched whistle is used as a source of sound and a sensitive flame as a detector, then it is found **Sound shadows.** that an obstacle prevents the sound reaching the flame if it is cut by the straight line joining the whistle to the base of the flame. The obstacle casts a sound shadow just like the shadows produced by a source of light. With sounds of lower frequency shadows can also be obtained provided the size of the obstacle is increased in the same proportion as the wave length of the sound.

For example, a cannon fired on one side of a hill may not be audible on the other side although it can be heard at much greater distances in other directions. To get a shadow the obstacle must be large compared with the wave length of the sound. The theory of the formation of shadows will be more fully discussed in the chapters on light, which, like sound, is a form of wave motion.

If a point moves with a simple harmonic motion parallel to a fixed line and also at the same time with another **Composition of two vibrations at right angles.** simple harmonic motion parallel to a perpendicular line, the curve it describes is the result of the composition of two simple harmonic motions at right angles to each other.

Let the two motions be represented by the equations

$$y = A \cos 2\pi n t$$

and $$x = A' \cos (2\pi n' t + \alpha);$$

x and y may then be taken to be the coordinates of the point, and the relation between x and y got by eliminating t from these two equations is the equation of the curve described by the point. If the two perpendicular motions have equal amplitudes and frequencies and $\alpha = 0$, we get

$$y = A \cos 2\pi n t$$

and $$x = A \cos 2\pi n t,$$

so that $x = y$. In this case the curve described is simply a line inclined at 45° to the directions of the two vibrations.

If $A = A'$, $n = n'$ and $\alpha = \pi/2$, we get

$$y = A \cos 2\pi n t,$$

$$x = A \cos (2\pi n t + \pi/2) = - A \sin 2\pi n t.$$

Hence $$x^2 + y^2 = A^2 (\cos^2 2\pi n t + \sin^2 2\pi n t),$$

so that $x^2 + y^2 = A^2$. In this case therefore the curve described is a circle of radius A.

The result of the composition of two simple harmonic motions of equal amplitudes, at right angles, can be obtained graphically as shown in Fig. 23. We begin by taking a line AOB bisected at O and describe a circle with centre O and radius OB. Then draw a diameter COD perpendicular to AOB. Divide the arc CB into a number of equal parts, say six, and divide each of the arcs CA, AD and DB in the same way. Through all the points dividing the circle into 24 equal parts draw lines parallel to AB and also draw lines parallel to CD as shown. In this way we get a square divided into a number of rectangles. The distances from one vertical line to the next are equal to the distances described in equal times by a point moving with a simple harmonic motion parallel to AB of amplitude OB. In the same way the distances between successive horizontal lines are equal to the distances described in equal time intervals by a point moving with a simple harmonic motion parallel to CD of amplitude OC.

Suppose we wish to get a curve described by a point moving with two simple harmonic motions at right angles of equal frequencies and amplitudes. Let the point start at the intersection of any horizontal with any vertical line as at P. It will then move to the next vertical line in the same time that it moves to the next horizontal line because the two frequencies are equal. It will

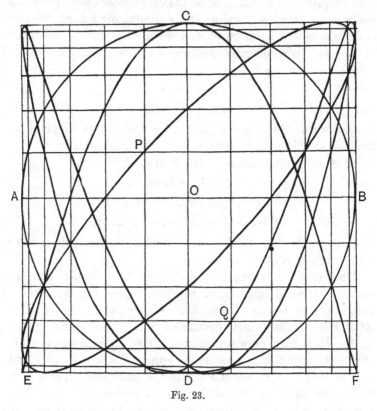

Fig. 23.

therefore move round the curve drawn through P which always passes from one corner of a rectangle to the opposite corner. If we suppose the point starts at A, then in the same way we see that it will go round the circle $ACBD$. If it starts at O it will move backwards and forwards along a diagonal of the square.

Suppose we wish to get a curve described by a point moving with two simple harmonic motions at right angles of equal

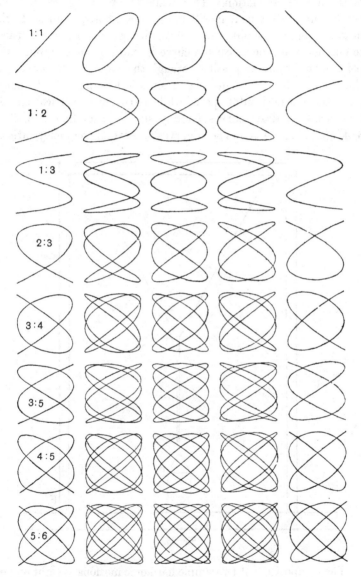

Fig. 24.

amplitudes but with the period of the vertical motion half that of the horizontal motion. Then while the point moves from one vertical line to the next it will move from a horizontal line to the next horizontal line but one. That is it covers two vertical spaces to one horizontal one. Such a curve is shown passing through the point Q and another passing through the points E, C and F. If the ratio of the two frequencies desired is n to m, then the curve must cover n horizontal spaces while it covers m vertical spaces. It is easy to draw curves for any simple ratio like $1:3$, $2:3$, $3:4$, etc. Fig. 24 shows such curves for several frequency ratios.

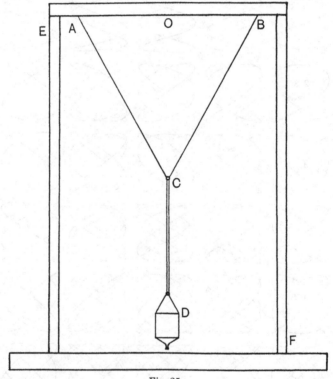

Fig. 25.

The composition of two simple harmonic motions at right angles may be obtained experimentally in a number of different ways. One of the simplest, known as Blackburn's pendulum, is shown in

Fig. 25. A stand EF supports a horizontal bar AB. Two strings ACD and BCD support a can D which serves as the bob of the pendulum. The strings are held together at C by a clip which can be moved up and down the strings. If the bob D oscillates in the plane of the paper, the point C remains fixed and the period is given by $T = 2\pi \sqrt{\dfrac{CD}{g}}$; but if the bob oscillates in a plane perpendicular to the plane of the paper, then the strings AC and BC swing with CD so that $T' = 2\pi \sqrt{\dfrac{OD}{g}}$. If we wish to observe the curves described when the periods are in the ratio $n : m$ we adjust the clip so that

$$\frac{n}{m} = \frac{T}{T'} = \sqrt{\frac{CD}{OD}} \quad \text{or} \quad \frac{CD}{OD} = \frac{n^2}{m^2}.$$

The can D is usually made with a small hole at its lowest point. If it is filled with sand, then a small stream of sand runs out and the curves described by the bob are traced out in sand on the base of the instrument.

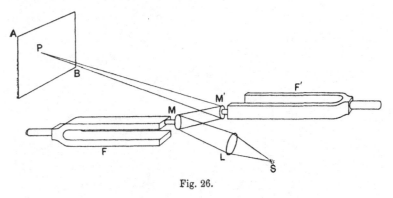

Fig. 26.

The composition of the simple harmonic motions of the prongs of two tuning forks vibrating in perpendicular directions can be demonstrated with the apparatus shown in Fig. 26. F and F' are the two forks, which each have small plane mirrors M and M' fixed to one prong. Light from a small source S, such as a strongly illuminated pin hole in a metal sheet, passes through a lens L and is reflected first by M and then by M' on to a screen AB. The lens L is adjusted so that it forms an image of S at P on the screen. If the fork F' is set vibrating P oscillates in a horizontal line, and if the fork F is set vibrating P moves in a vertical line. If both F and F' are set vibrating the motion of P is the result of compounding together the two perpendicular simple harmonic motions. If the two forks have exactly equal frequencies the curve described by P is in general an ellipse. If the frequencies are very nearly but not exactly equal, the shape of the ellipse slowly changes and passes successively through shapes like those shown in Fig. 24 for the ratio $1:1$. The shape of the ellipse depends on the phase difference of the two vibrations, which varies slowly when the frequencies are not exactly equal. With forks having frequencies nearly but not exactly in a simple ratio like $1:2$ or $2:3$ we get the corresponding curves which slowly change in shape as the phase difference varies. This arrangement was first described by Lissajous and the curves obtained are sometimes called Lissajous' figures.

CHAPTER VI

RESONANCE

When a tuning fork or a pendulum is set vibrating it oscillates with a definite frequency. This frequency is sometimes called the frequency of free vibration, and the time of one vibration is called the period of free vibration or the free period. The frequency of free vibration is obtained when the tuning fork or other body is allowed to vibrate freely with no external forces acting on it. It is possible to make a body vibrate with a frequency different from

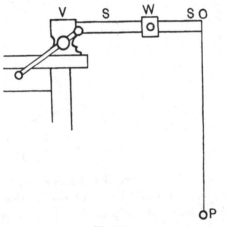

Fig. 27.

its natural frequency of free vibration by allowing suitable forces to act on it. Such a vibration is called a forced vibration. As an example of a forced vibration consider the case of a simple pendulum suspended from the end of a horizontal spring which can vibrate in a horizontal plane. Such an arrangement is shown in Fig. 27. V is a vice fixed to a table, in which a steel or hard brass strip SS is firmly clamped. This strip may be 2 cms. broad, 0·5 mm. thick

18

and about 50 cms. long. A weight W can slide along the strip and be clamped on it in any desired position. A simple pendulum OP, having a very light bob, is hung from the free end of the spring. If the spring is pulled to one side and let go it vibrates in a horizontal plane, so that O moves with a simple harmonic motion perpendicular to the plane of the paper. Let the period of this vibration be T' and let the length of the pendulum be l, so that if O was a fixed point the free period T of the pendulum would be given by $T = 2\pi \sqrt{l/g}$. First suppose T is greater than T', then it is found that the pendulum swings as shown in Fig. 28. The end of the spring oscillates between O and O' and the pendulum bob between P' and P with the period T'. The point A in the string remains at rest. The direction of the displacement of P

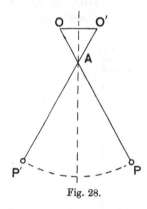

Fig. 28.

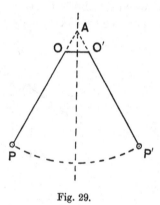

Fig. 29.

is opposite to that of O. Since A remains at rest the pendulum swings like a pendulum of length $AP = l'$, so that $T' = 2\pi \sqrt{l'/g}$. The amplitude of swing of the pendulum bob is equal to

$$\tfrac{1}{2}OO' \frac{AP}{AO} = a \frac{l'}{l-l'},$$

where $a = \tfrac{1}{2}OO'$ is the amplitude of the vibration of the end of the spring.

Since $\qquad T = 2\pi \sqrt{l/g}$ and $T' = 2\pi \sqrt{l'/g}$,

we have $\qquad a \dfrac{l'}{l-l'} = a \dfrac{T'^2}{T^2 - T'^2}.$

We see from this that if T' becomes nearly equal to T the amplitude of swing of the pendulum bob becomes very large compared with the amplitude of the end of the spring.

Suppose now that T' is made greater than T by moving the sliding weight nearer to the end of the spring. The motion of the pendulum in this case is shown in Fig. 29. The displacement of the bob P is now in the same direction as that of the spring O, and if PO is produced upwards it always passes through a fixed point A. We have

$$\tfrac{1}{2}PP' = \tfrac{1}{2}OO' \times \frac{AP}{AO} = \frac{al'}{l'-l} = \frac{aT'^2}{T'^2-T^2},$$

where $\qquad a = \tfrac{1}{2}OO', \quad l = OP, \quad l' = AP,$

$T = 2\pi\sqrt{l/g}$ and $T' = 2\pi\sqrt{l'/g}$ as before.

In this case also the amplitude of swing of the pendulum becomes very large when T' is reduced nearly to the value T.

The large amplitude of vibration produced when the two periods coincide is an example of what is called resonance. Resonance takes place when a periodic force acts on a body, and the period of this applied force is equal or nearly equal to the period of free vibration of the body. The body is then set vibrating with a period equal to the period of the force and vibrates with a large amplitude.

Consider the case of a body of mass M, which can move along a straight line, acted on by a force F proportional to the displacement of the body from a fixed point in the line and directed towards the point. Let $F = -\mu x$ where μ is a constant. We know that if the body is displaced and then let go it oscillates with a simple harmonic motion of period T and frequency n given by

$$T = \frac{1}{n} = 2\pi\sqrt{M/\mu}.$$

Now suppose that another force F' besides F acts on this body, and let F' be a periodic force given by the equation

$$F' = P\cos 2\pi n't,$$

where t denotes the time, P is a constant and n' is the number of times per second that F' oscillates between the limits P and $-P$ and back again to P. Under the action of this periodic force the body will execute a forced vibration of frequency n'. Let its displacement x be given by the equation

$$x = A \cos 2\pi n' t.$$

The resultant force on the body at any time t is

$$F' - \mu x = P \cos 2\pi n' t - \mu x,$$

so that its acceleration a is given by the equation

$$P \cos 2\pi n' t - \mu x = Ma,$$

but
$$\cos 2\pi n' t = x/A,$$

so that
$$\frac{P - \mu A}{A} x = Ma.$$

Thus Ma which is equal to the resultant force on the body is proportional to x, and the period T' of vibration of the body must be given by the equation

$$T' = \frac{1}{n'} = 2\pi \sqrt{\frac{MA}{\mu A - P}},$$

which gives
$$A = \frac{P}{\mu - 4\pi^2 n'^2 M}.$$

But
$$\mu = 4\pi^2 M n^2,$$

so that finally
$$A = \frac{P}{4\pi^2 M (n^2 - n'^2)}.$$

When the frequency n of free vibration of the body is greater than n', the frequency of the applied force, then at any instant the displacement of the body is in the same direction as the force F' for $n^2 - n'^2$ is positive, so that A and P both have the same sign. When n' is greater than n, A and P are of opposite signs, so that the displacement of the body is in the opposite direction to the periodic force F'. When n' is nearly equal to n, A becomes very large compared with P. If $n' = n$, then A becomes infinite theoretically, but in practice owing to friction and other causes the amplitude cannot rise above a certain finite value. In Fig. 30 the variation of the amplitude A with the frequency n' of the applied force F' is shown graphically. When n' is very small A has a constant small positive value independent of n'. As n' approaches the free frequency n, A rises and becomes $+\infty$ at

$n' = n$. When n' is greater than n and very large, A becomes zero. If n' is diminished, as it approaches n, A becomes appreciable and is negative. When n' becomes equal to n, $A = -\infty$. Thus at $n' = n$, A changes from $+\infty$ to $-\infty$. In practice A of course does not rise to infinity but to a large value represented at B, and then changes to an equal value of opposite sign represented at C.

The experiment described at the beginning of this chapter serves as a good illustration of the theory just given. The string exerts a periodic force on the pendulum bob. This experiment

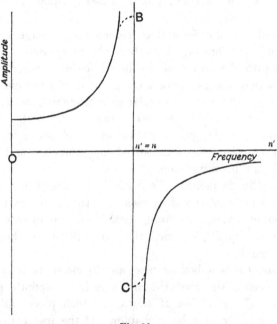

Fig. 30.

may be done more simply by holding the string in the hand and moving it backwards and forwards with different frequencies. A striking experiment illustrating resonance can be shown with a wheel mounted in bearings. If a stiff spring is attached to one of the bearings and then the wheel set rapidly rotating, as the wheel gradually slows down, a time comes when the period of revolution of the wheel is equal to the period of vibration of the spring. When this occurs the spring begins to vibrate violently.

The wheel shakes its bearings slightly with a period equal to its period of revolution.

If a weight is hung up by a spiral spring, it can oscillate up and down with a definite period. If two equal springs and weights are hung from the same support, then if one of them is set vibrating it exerts a periodic force on the support so that this moves slightly if it is not too rigid. The motion of the support causes a small periodic force to act on the other spring and weight which gradually sets it vibrating with a large amplitude, because the period of the force acting on it is nearly equal to its own free period.

A small periodic force if continued long enough may set a very large body vibrating violently if the free period of the body coincides with the period of the force. Soldiers when marching over a bridge always break step, because if they did not the periodic forces due to their regular marching might set the bridge vibrating dangerously if the free period of the bridge happened to be equal to the period of vibration of the soldiers' legs. Large bridges have been destroyed by soldiers marching over them without breaking step.

When the frequencies lie within the limits of audibility resonance can be observed by means of the sound produced. If the handle of a vibrating tuning fork is allowed to touch a body with a nearly equal free period, the body will be set vibrating and give out sound.

A resonator is a box or pipe usually closed at one end and made of such a size that the air in it has a definite period of vibration. The vibration of the air in such pipes will be discussed more fully in a later chapter. If the free period of the air in a resonator is equal to the free period of a tuning fork, then when the fork is sounded near the resonator the air in the resonator is set vibrating. If the fork is mounted on the top of the resonator, then sounding the fork sets the air in the resonator vibrating strongly, and a loud sound is produced. Tuning forks mounted on resonators are often useful in experiments on sound. They give out a strong sound of definite frequency. Fig. 31 shows a fork mounted on a wooden resonator.

If two forks of exactly equal frequencies both mounted on

their resonators are in the same room, then if one of them is sounded, some of the sound from it falls on the other and sets it vibrating. A small hard ball hung up by a thread may be allowed just to touch the prongs of the second fork. When the first fork is sounded the second fork knocks the ball away. With accurately adjusted forks this experiment succeeds even if the forks are

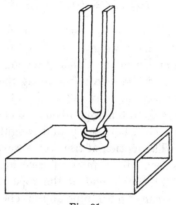

Fig. 31.

30 feet or more apart. If a small piece of wax is then stuck to one of the prongs of either fork, the two periods are then no longer equal, and sounding one fork does not set the other vibrating. Several other examples of resonance will be described in the next chapter.

CHAPTER VII

VIBRATION OF STRINGS

If a long flexible cord is fixed at one end and the other end is passed over a pulley and a weight attached to it, then on pulling the cord near the pulley to one side and letting it go a transverse disturbance or wave can be seen to run along the cord. When the wave reaches the fixed end it is reflected and comes back in the opposite direction. A cotton rope about 0·5 cm. in diameter and 20 metres long stretched by a ten pound weight may be used for this experiment. The motion of transverse waves along a flexible cord can also be shown with a piece of rope hung vertically from a fixed point. If the lower end of the rope is suddenly moved to one side, a wave runs up the rope and is reflected at the top.

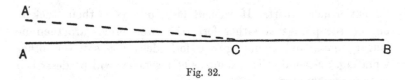

Fig. 32.

The velocity with which a transverse wave travels along a flexible string can be easily calculated. Let the tension in the string be P and the mass of the string per unit length be m. In Fig. 32, AB is the string in its undisturbed position. Suppose now that A is moved upwards perpendicularly to AB with uniform velocity v'. After a time τ, A will have reached A' and $AA' = v'\tau$. If the velocity with which a transverse wave moves along the string is v, then the string will have begun to move upwards like A for a distance $AC = v\tau$ but beyond C it will still be undisturbed. The string between A and C will be moving up with the velocity v' communicated to it by the motion of A. The momentum given to the string in the time τ is therefore equal to $mv\tau v'$. The force required to move A with velocity v' is equal and opposite to the component of the tension T in $A'C$ along $A'A$. This component is equal to

$$P \times \frac{A'A}{A'C} = P \frac{v'}{v}.$$

But this force must be equal to the momentum communicated to the string per second so that

$$P \frac{v'}{v} = mvv',$$

or

$$v = \sqrt{\frac{P}{m}}.$$

This result may also be obtained in another way as follows. Imagine the string to be passed through a smooth tube CD part of which between A and B (Fig. 33) is bent into a curve of any

Fig. 33.

shape. The string presses against the tube with a force equal to P/r per unit length, where r is the radius of curvature of the tube at the point considered. To prove this consider two points N and N'

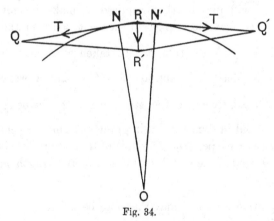

Fig. 34.

very near together on the string (Fig. 34). Draw NO and $N'O$ perpendicular to the string. Then $ON = r$. Let RQ and RQ' represent the forces exerted by the rest of the string on the part of it between N and N'. These forces are both equal to the tension P in the string. Complete the parallelogram $QRQ'R'$ so that RR' represents the resultant force on NN'. The triangle RQR' is similar to the triangle NON' so that

$$\frac{RR'}{RQ} = \frac{NN'}{NO}.$$

The resultant force is therefore equal to $P\dfrac{NN'}{NO}$. Let $N\hat{O}N' = \theta$

so that $NN' = r\theta$ and $P\dfrac{NN'}{NO} = P\dfrac{r\theta}{r} = P\theta.$

The force per unit length on the string is therefore equal to

$$\frac{P\theta}{r\theta} = \frac{P}{r},$$

and is directed towards O.

If the string slides through the tube with a velocity v, the force per unit length on it, required to keep it on its curved path, is equal to mv^2/r since m is the mass of unit length. If v is increased until

$$\frac{mv^2}{r} = \frac{P}{r} \quad \text{or} \quad v = \sqrt{\frac{P}{m}},$$

then the resultant force P/r on any portion of the string due to the tension in it will just be that required to make it move in its curved path, so that the string will then not press on the tube at all. When $v = \sqrt{\dfrac{P}{m}}$ the tube can be taken away and the string will retain its shape. It appears therefore that a wave in the string moves relatively to the string with the velocity $\sqrt{\dfrac{P}{m}}$. If P is expressed in dynes and m in grams per cm. the velocity will be expressed in cms. per sec. The unit of tension or force is MLT^{-2} where M, L and T denote the units of mass, length and time respectively.

The equation $v = \sqrt{\dfrac{P}{m}}$ may therefore be written

$$v\frac{L}{T} = \sqrt{\frac{PMLT^{-2}}{mML^{-1}}} = \sqrt{\frac{PL^2}{mT^2}} = \sqrt{\frac{P}{m}} \times \frac{L}{T}$$

which shows that both sides of it represent a velocity.

If an endless flexible cord ABC is hung over a pulley P rapidly driven by a motor as shown in Fig. 35, then the cord will retain the shape shown however rapidly the pulley is made to rotate. The tension in the cord due to its motion round the curved ends of its path is given by the equation $\dfrac{mv^2}{r} = \dfrac{P}{r}$ where v is the velocity of the cord so that $P = mv^2$, and therefore for any speed the velocity of a wave along the cord is equal to the velocity of the cord. Consequently the cord retains its shape like the string moving with velocity $\sqrt{\dfrac{P}{m}}$ when the curved tube surrounding it is removed. The curved part DBE moves along the cord with a velocity $\sqrt{\dfrac{P}{m}}$ which is equal and opposite

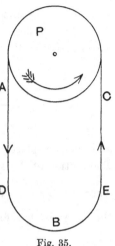

Fig. 35.

to the velocity of the cord. The position of the curve DBE consequently remains fixed.

When a transverse wave moving along a string comes to a fixed point it is reflected so that it returns along its path. At the fixed point the sum of the two displacements due to the incident and the reflected waves must be zero, so that it is clear that the displacements in the reflected wave must be equal and opposite to those in the incident wave.

In musical instruments like the violin, banjo and piano the sound is produced by stretched strings or wires, fixed at each end, which are made to vibrate transversely. If a stretched string fixed at each end is struck with a hammer or pulled to one side and then let go, two transverse waves are produced in the string which move away from the place where it was struck in opposite directions with the velocity $\sqrt{\dfrac{P}{m}}$. These waves are reflected at the fixed ends and their displacements are reversed in direction at each reflexion. Let the length AB (Fig. 36) of the string be l, and let it be struck at a point P distant x from one end A. One of the waves starts from P towards A and is reflected there. It then goes to B and is again reflected and moves back towards P. After the two reflexions the displacements in the wave are in the same direction as at the start. When the wave arrives at P from B it has gone a distance $PA + AB + BP = 2l$. The other wave goes from P to B, then from B to A and then from A to P so that both waves arrive at P at the same time after travelling the distance $2l$

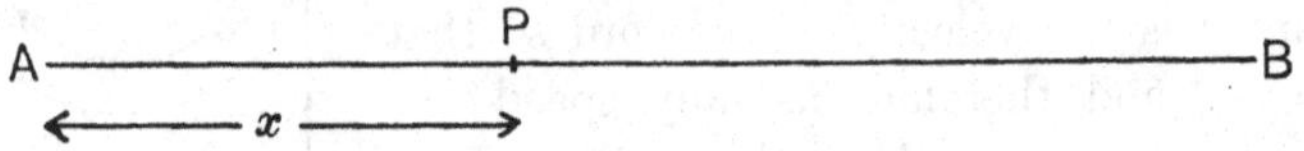

A ————————————————————•———————————————————————— B

← —————— x —————— →

Fig. 36.

and their displacements are then the same as that at the start. The motion of the string therefore repeats itself after the time $2l/v$ where $v = \sqrt{\dfrac{P}{m}}$. The fundamental note in the sound emitted by the string is therefore of frequency

$$n = \frac{v}{2l} = \frac{1}{2l}\sqrt{\frac{P}{m}}.$$

If the middle point of the string is held fixed, each half can vibrate with frequency $\dfrac{1}{l}\sqrt{\dfrac{P}{m}}$ just like the whole string. If the string is held fixed at a series of equidistant points which divide it into parts each of length l/k where k is a whole number, then each part has a frequency of vibration $\dfrac{k}{2l}\sqrt{\dfrac{P}{m}}$. When the string is only fixed at each end it can vibrate with any or all of the frequencies represented by $n = \dfrac{k}{2l}\sqrt{\dfrac{P}{m}}$ where k is equal to 1 or 2 or 3 or any other whole number. For example, if we hold the string at a point one-third of l from one end and pluck the string half way between this point and the nearer end, this part of the string is set vibrating with the fundamental frequency $\dfrac{3}{2l}\sqrt{\dfrac{P}{m}}$. The other two equal parts have equal frequencies so that they are set vibrating by the small periodic disturbances which get from the first third of the string to the rest. Thus all three thirds of the string are set vibrating but the two dividing points remain at rest. When a string is struck or plucked it is usually set vibrating so that many of the possible vibrations with frequencies

$$\frac{v}{2l}, \quad \frac{2v}{2l}, \quad \frac{3v}{2l}, \quad \frac{4v}{2l} \quad \text{etc.}$$

are present. Thus we get a fundamental of frequency $v/2l$ and a series of harmonics with frequencies which are exact multiples of the fundamental frequency.

The vibration of stretched strings can be studied with the apparatus shown in Fig. 37, which is called a monochord. BB is a wooden box about 120 cms. long. A wire or string PWQ is stretched along the top of the box over two knife edges at P and Q. The end of the wire near P is attached to a spring balance S which serves to measure the tension in the wire. The other end is wrapped round a conical plug R which fits rather tightly into a hole in a wooden block fastened to the end of the box. A moveable knife edge W slightly higher than those at P and Q can

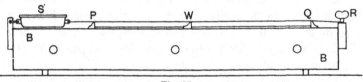

Fig. 37.

be put under the wire at any desired point. A millimetre scale fixed to the box parallel to the wire between P and Q serves to measure the length of any part of the wire. If the wire is struck or plucked it vibrates and sets the box vibrating so that an easily audible sound is produced. The wire alone would produce very little sound because its surface is so small. To verify the formula

$n = \dfrac{1}{2l} \sqrt{\dfrac{P}{m}}$ we may vary the length and tension in the wire until

it gives a note of the same frequency as a tuning fork of known frequency. The length is varied by moving the bridge W and the tension by turning the plug R. The wire presses against the knife edges so that they hold it fixed when it vibrates up and down. A good way to tell when the wire and fork have equal frequencies is by means of resonance. If the frequencies are equal and the handle of the fork is held against the box when the fork is sounding, the wire will be made to vibrate strongly. A small piece of paper bent into a **V** shape may be put on the wire near the middle of the part which vibrates and if the frequencies are equal touching the box with the fork will make the paper jump off.

In this way it can be shown that when the frequency is constant l is proportional to $\sqrt{P}$. By using different forks it can be shown that n is proportional to $\sqrt{P}$ when l is constant and inversely proportional to l when P is constant. By using wires of different thicknesses and materials and known masses it can be shown that when l and P are kept constant n varies inversely as $\sqrt{m}$ and does not depend on the material of which the wire is made. Finally if P, l and m are all measured $\dfrac{1}{2l}\sqrt{\dfrac{P}{m}}$ can be calculated and will be found to agree approximately with the frequency observed.

The vibration of the wire with frequencies higher than its fundamental can be easily shown. Let the bridge W be put at a point 20 cms. from one end of the wire, the whole length l of which is 100 cms. Then place paper riders at points 30, 40, 50, 60, 70, 80 and 90 cms. from the same end. If the wire is now plucked half way between the bridge and the end near it, the riders at 30, 50, 70 and 90 cms. will jump off while those at 40, 60, and 80 cms. will not be disturbed. Thus the wire vibrates in five equal sections and the points 20, 40, 60 and 80 cms. from the end remain at rest. The note emitted can be recognised as having a frequency five times that of the fundamental note given by the whole wire. If the bridge is put 25 or 33·3 cms. from one end, the wire can be shown in the same way to vibrate in 4 or 3 sections with frequencies 4 or 3 times that of the fundamental. The points which remain at rest, when the wire is vibrating with one of its higher frequencies, are called the nodes.

The modes of vibration of a stretched string can be very beautifully shown by a method due to Melde. The apparatus used in Melde's experiment is shown in Fig. 38. F is a large tuning fork mounted vertically. A string is fastened to one of the prongs of the fork at A and passes over a pulley P to an adjustable weight W. If the weight W and the distance AP are adjusted so

that one of the possible periods of vibration of the string is equal
to half the period of the fork, then when the fork is set vibrating
the string will be made to vibrate with a large amplitude which
can be clearly seen. In Fig. 38 the string is shown vibrating in
three sections with nodes at A, B, C and P. By adjusting W and
the length AP the string can be made to vibrate in different numbers
of sections. For example if W is diminished to one-quarter of its
value the string vibrates in six sections instead of in three and so
on. If the fork is turned through a right angle so that it vibrates
in a plane perpendicular to the string, then the frequency of the
string is doubled so that the length of a section is halved if the

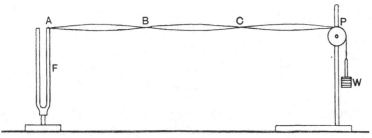

Fig. 38.

tension in the string is not changed. The number of sections can
be changed from two to four or from three to six by turning the
fork through a right angle. If S denotes the length of a section,
the frequency of the string is $\dfrac{1}{2S}\sqrt{\dfrac{P}{m}}$. This formula can be
verified, if the frequency of the fork is known, by measuring S, P
and m.

The vibration of violin strings when bowed was investigated
by Helmholtz with an instrument called a vibration microscope
shown in Fig. 39. S is the violin string the length of which is
supposed to be perpendicular to the plane of the paper. F is
a tuning fork one prong of which carries the object glass O of
a microscope M the eye-piece of which is at I. The fork F
can be kept vibrating electrically so that the objective vibrates
perpendicularly to the plane of the paper and parallel to the length
of the string S. A small silvered bead is fastened to the string at
S and the bead is illuminated with an electric lamp or other

convenient source of light. The microscope is focused on the
bead, so that when the string is at rest and the fork vibrating the
image of the bead as seen in the microscope vibrates parallel to
the string and so looks like a straight line.   If the string is bowed

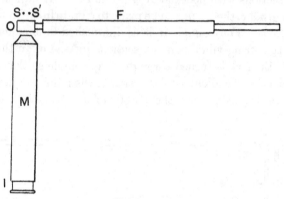

Fig. 39.

so that it vibrates perpendicularly to the axis of the microscope
between S and S', the motion of the image of the bead is due to
the composition of the simple harmonic motion of the fork and the
perpendicular vibration of the string.   If the tension of the
string is adjusted so that its fundamental frequency is equal
to the frequency of the fork, the image of the bead in the
microscope appears to be a curve like one of those shown in
Fig. 40.

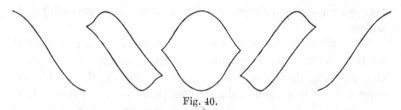

Fig. 40.

Curves like these can be obtained in the following way.
Describe a circle on AB (Fig. 41) as diameter and draw a perpendicular diameter CD. Divide each of the four arcs CB, BD, DA
and AC into a number of equal parts, say 6. Through the dividing
points draw horizontal lines. Then divide AB into 12 equal parts

and through the dividing points draw vertical lines. If curves are drawn so that they join opposite corners of successive rectangles we get curves as shown, which are like those seen in the vibration microscope when the string is bowed. These curves represent the composition of a simple harmonic motion with a perpendicular vibration in which the velocity is constant in magnitude but changes sign at the end of each swing.

The motion of a point on a violin string while it is bowed is therefore a uniform velocity reversed in direction at regular intervals. The velocity in one direction may be different from that in the opposite direction.

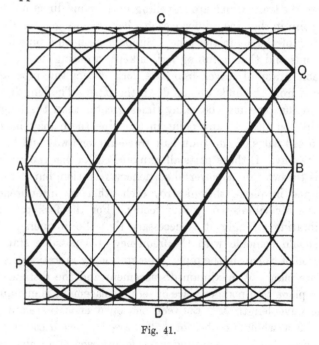

Fig. 41.

REFERENCES

The Dynamical Theory of Sound, Horace Lamb.
Sound, Poynting and J. J. Thomson.

CHAPTER VIII

VIBRATION OF AIR IN OPEN AND CLOSED PIPES

WE have seen in Chapter V that a train of waves travelling along through the air in a pipe is reflected from either a closed or an open end of the pipe. A train of waves in a pipe is therefore reflected up and down the pipe between the ends. If the different parts of the train, which are travelling in the same direction, at any fixed point in the pipe reinforce each other, then we get a stationary vibration of the air in the pipe. Suppose we have a pipe open at both ends and that a tuning fork is kept vibrating near one end. A feeble train of waves from the fork enters the pipe and is reflected from the ends up and down the pipe. The length of the train in the pipe may be many times greater than the length of the pipe, so that at any point in the pipe there may be many superposed parts of the train half travelling one way and half the opposite way. If these parts all reinforce each other the air in the pipe is thrown into a powerful stationary vibration, but if they do not agree in phase they destroy each other by interference so that the air vibrates only very feebly. For the fork to set the air vibrating strongly it is necessary that the frequency of the fork should coincide with the frequency of a possible stationary vibration of the air in the pipe. In a stationary vibration the open ends are planes of maximum displacement and the distance from such a plane to a node or place of zero displacement is one-quarter of the wave length λ. Between the open ends therefore there must be a whole number of half wave lengths because planes of maximum displacement and nodes follow each other alternately.

If l is the length of the pipe we have therefore

$$\frac{l}{m} = \frac{\lambda}{2},$$

where m is any whole number.

Hence $\qquad\qquad \lambda = 2l/m.$

Fig. 42 shows the variation of the amplitude of vibration along a pipe open at both ends for different values of m.

The vertical distance between the two curves is supposed to represent the amplitude of the vibration. The direction of the vibration of course is always parallel to the length of the pipe.

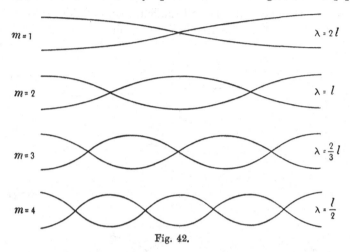

$m = 1$ $\qquad \lambda = 2l$

$m = 2$ $\qquad \lambda = l$

$m = 3$ $\qquad \lambda = \frac{2}{3}l$

$m = 4$ $\qquad \lambda = \frac{l}{2}$

Fig. 42.

Where the two curves cross each other the amplitude is zero so that the vertical planes through these points are nodes.

In the case of a pipe closed at one end and open at the other a stationary vibration in the pipe must have a plane of zero displacement at the closed end and a plane of maximum displacement at the open end. The length l of the pipe must therefore be equal to an odd number of quarter wave lengths. Hence

$$l = (2m + 1)\frac{\lambda}{4},$$

or $\qquad\qquad \lambda = 4l/(2m + 1),$

where m denotes any whole number. Fig. 43 shows the variation of the amplitude of vibration along pipes closed at one end. The closed end is supposed to be on the left.

Since $v = n\lambda$, where v is the velocity of sound in air and n the frequency of a train of wave length λ, we have for pipes open at both ends

$$n = \frac{v}{\lambda} = \frac{vm}{2l},$$

where $m = 1, 2, 3, 4$, etc.; and for pipes closed at one end

$$n = \frac{v}{\lambda} = \frac{v}{4l}(2m + 1),$$

where $m = 1, 2, 3$, etc.

The fundamental frequency of open pipes is $v/2l$ and the higher possible frequencies include all the harmonics of this fundamental note. The fundamental frequency of a pipe closed at one end is

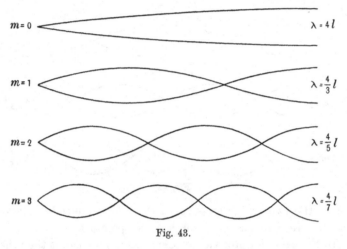

Fig. 43.

$v/4l$ or half that of an open pipe of the same length. The higher possible frequencies include only the harmonics having frequencies which are odd multiples of that of the fundamental note.

The air in the pipes used in organs and other musical instruments is sometimes made to vibrate by blowing a current of air from a slit across one of the open ends of the pipe. Such an organ pipe is shown in Fig. 44.

The air enters at A and blows through a narrow slit towards a sharp edge on the other side of the opening at B. The end C may be either open or closed. The air in the pipe between B and C is thrown into a state of stationary vibration which includes all the possible states, so that the pipe produces its fundamental note together with a series of harmonics. If the end C is open we get harmonics with frequencies $2, 3, 4, 5$, etc. times that of the fundamental, and if C is closed we get a fundamental of half the frequency and harmonics with frequencies 3, 5, 7, 9, etc. times that of the fundamental. The quality of the sound emitted by an open pipe is quite different from that of a pipe closed at one end. The fact that there may be a node near the middle of an open pipe can be shown with a vertical glass pipe by means of a light paper box containing some sand. The box is hung up by a thread and lowered into the pipe. Near the top or bottom of the pipe the paper vibrates up and down and shakes the sand about, while near the middle of the pipe it remains at rest. The box tends to prevent the air vibrating so that its presence at any point in the pipe favours the production of those stationary vibrations which have

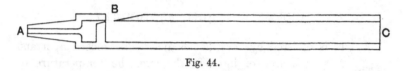

Fig. 44.

nodes near where it is and more or less completely stops the others. The box will always show a node near the middle of the pipe and also usually nodes half way between the middle and either end corresponding to the first harmonic.

It is found that if an open organ pipe is blown with air at more than a certain pressure it ceases to produce its fundamental note and gives only the harmonics, so that the fundamental frequency of the sound emitted is doubled. When it is sounding in this way the paper box and sand may show the two nodes corresponding to the first harmonic and no node at the middle of the pipe, but usually the presence of the box near the middle is sufficient to start the fundamental vibration and more or less completely to stop the first harmonic.

PART IV

LIGHT

CHAPTER I

SOURCES OF LIGHT, PHOTOMETRY

LIGHT is emitted by very hot bodies. The most important source of light is the sun, the temperature of which is estimated to be about 6000°C. If the temperature of a solid body in a dark room is gradually raised, it begins to emit light and so becomes visible at about 400°C. At this temperature it appears dull red. At about 1000°C. a solid body is bright red hot, and at 1500°C. it is white hot, and emits a bright light. Artificial light is almost always obtained from hot solid bodies. Candle and coal gas flames contain an immense number of minute particles of solid carbon which at the temperature of the flame, about 1500°C., emit nearly white light. Incandescent electric lamps contain a thin wire made of carbon, tungsten or some other solid body which is kept at a very high temperature by passing a current of electricity through it. Mercury arc lamps consist of a glass or quartz tube in which a current of electricity is passed through mercury vapour. In these lamps the light is emitted by the mercury vapour, which is not very hot, so that they form an exception to the rule that light is usually obtained from hot solid bodies. Ordinary electric arc lamps consist of two carbon or magnetite rods with their ends near together. A current of electricity is passed from one to the other across the gap between them, and the ends of the rods become very hot and emit light.

Sources of Light.

A small source of light in an otherwise dark room casts

Shadows.   shadows of bodies on the walls of the room. A straight line drawn from any point on the boundary of a shadow to the source is found to touch the surface of the body producing the shadow. This may be proved with the apparatus shown in Fig. 1. S is a small incandescent electric lamp, B a body of any shape, and O a white screen. If one end of a straight rod is put at a point like P on the boundary of the shadow of B on the screen O and the other end at S, then the rod will just touch the surface of the body B at the point A. This shows that the light from S travels along straight lines through the space surrounding S. If several metal sheets each containing

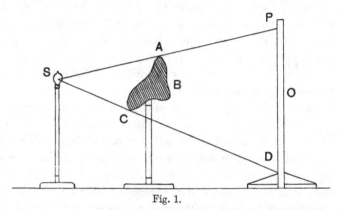

Fig. 1.

a small hole are put up between S and O, then if the holes and S all lie in the same straight line so that the straight rod can be passed through the holes, the light from S will pass through the holes and illuminate a spot on the screen. We conclude that light travels through air in straight lines.

If the boundary of the shadow of the body B on the screen O is carefully examined it is found that it is not perfectly sharp and that there is some illumination of the screen slightly below the straight line SAP. If straight lines are drawn from S so as to touch the body B, and are produced to the screen, these lines all meet the screen along a curve which is called the boundary of the geometrical shadow of B. It is found that the shadow actually obtained is slightly smaller than the geometrical shadow and not perfectly sharply defined. It appears therefore that the light which passes close to B deviates slightly from a perfectly straight path. This deviation is called diffraction and will be discussed in a later chapter.

For many purposes light may be regarded as travelling in straight lines, because the deviation of the light which passes close to the surface of a body is so small that it can usually be neglected and the light which does not pass close to the surface is not deviated at all. It is found that light travels through a vacuum just as it does through air. For example, the bulbs of incandescent electric lamps are very perfectly exhausted of air, so that the light has to pass across a vacuum before it can get out of the bulb. Substances like glass and water which allow light to pass through them are said to be transparent, while substances like metals through which light does not pass are said to be opaque. A vacuum is perfectly transparent, but all substances, even air and pure water, absorb some light, so that they are not perfectly transparent.

Rays. A line drawn from a source of light so that it everywhere coincides with the direction in which the light is travelling is called a ray of light. The rays from a small source in air or a vacuum are straight lines radiating out from the source in all directions. If the light from the small source falls on an opaque screen with a small hole in it, then the narrow beam of light which passes through the hole will follow the path of a ray of the light. If a white screen is put up perpendicular to the narrow beam, the centre of the bright spot on it marks the position of the ray which passed through the centre of the hole. Such a narrow beam of light is sometimes said to consist of a bundle of rays of light.

Nature of Light. The way in which a source of light produces shadows of bodies near it shows that light is something which starts from the source and moves from it in all directions. It is found that light travels through a vacuum with a velocity of 186,413 miles or 3×10^{10} cms. per second. It therefore takes 500 seconds to come from the sun to the earth, a distance of $1\cdot5 \times 10^{13}$ cms. When light falls on a black body it is absorbed and the body gets hotter, which shows that light has energy.

A hot body like an incandescent lamp filament can emit large quantities of light for a long time without any loss of weight. But energy has to be supplied to the filament while it is emitting light to keep its temperature constant. This energy is supplied by the electric current passing through the filament. Such a filament receives electric energy, which is converted into heat in the filament and partly emitted as light. The emission of light by the filament is analogous to the emission of sound by an electrically maintained tuning fork. The amount of matter in the fork remains constant but it receives electrical energy and emits some of it in the form of sound waves, which travel away from the fork through the surrounding air. The sound waves have energy; they are not a form of matter but merely a wave-motion in the air. Light is believed to be a wave-motion in a medium called the ether, which fills all space including the spaces between the atoms of material bodies. No way of removing the ether from any portion of space is known. What we call a vacuum is full of ether. The actions between bodies which seem to take place across empty space, like the attractions between the earth and other bodies or the attraction between a magnet and a piece of iron, are believed to be transmitted through the ether. The ether is believed to exist because such actions are found to take place, and because light travels through empty space and is found to have all the properties of a wave-motion travelling through a medium. The frequency and wave-length of trains of light waves can be determined experimentally, just as in the case of trains of sound waves in air. The methods by which this can be done will be discussed in later chapters.

An immense number of well-established facts can be explained by the theory that light is a form of wave-motion travelling through a medium which fills all space. No facts are known which are inconsistent with this theory, which is called the un-dulatory theory of light and is universally accepted among physicists. A small source of light is a centre of disturbances in the ether, which are propagated from it in all directions in the form of spherical light waves just as spherical sound waves spread out from a body vibrating in the air. Light waves are transverse waves, not longitudinal waves like sound waves in air.

If a light wave starts from a small source at a certain instant, then after a time t it will be a sphere of radius r such that $r = vt$, where v denotes the velocity of light. In a vacuum

$$v = 3 \times 10^{10} \frac{\text{cms.}}{\text{sec.}}.$$

In air and other gases the velocity of light is very slightly less than in a vacuum, but the difference is so small that it can be neglected in most cases.

The illuminating power of a source of light is taken to be proportional to the amount of energy in the form of light which it emits in unit time. The unit of illuminating power used for practical purposes is that of what is called a standard candle. At one time the illuminating power of a spermaceti candle $\frac{7}{8}$ inch in diameter burning away 120 grains per hour was used as the unit of illuminating power of sources of light. It is found that such candles do not always give the same amount of light, so that they do not provide a very satisfactory unit for exact work. The unit adopted by the International Congress of 1890 is one-twentieth part of the illuminating power of one square centimetre of liquid platinum at the melting point of platinum. This unit is nearly equal to one average standard spermaceti candle and is called a Decimal Candle. For making rough measurements ordinary paraffin candles may be used and assumed to be sources of unit illuminating power. The illuminating power of a source of light expressed in terms of that of the standard candle as unit is called the candle power of the source. The light waves from a small source spread out into the surrounding space in the form of spheres. The area of the surface of a sphere is proportional to the square of its radius, so the energy in the light waves passing through unit area in unit time from a small source is inversely proportional to the square of the distance from the source. This is true when none of the light is absorbed in the space surrounding the source. Light travels through air without appreciable absorption, so that the energy in the light waves from a candle or other small source in air passing through unit area in unit time varies inversely as the square of the distance of the area from the source.

The intensity of illumination of a surface is proportional to the amount of light falling on unit area in unit time. The unit intensity of illumination is that on a surface at unit distance from a unit source and perpendicular to the rays of light from the source. If we have a source of illuminating power K, that is to say of candle power K, then at a distance r it will produce an intensity of illumination I, on a surface perpendicular to the light rays from the source, given by the equation

$$I = \frac{K}{r^2}.$$

If $K = 1$ and $r = 1$, then $I = 1$.

The candle powers of two sources can be compared by means of instruments called photometers. A simple form of photometer is shown in Fig. 2. S is a white screen, R a vertical rod, C

Fig. 2.

a candle and E a source of light such as an incandescent electric lamp. The candle and lamp are arranged so that the shadows of the rod R which they produce on the screen S are close together but do not overlap. The experiment should be done in a room with blackened walls, into which no light can get from outside. The shadow produced by the rod and candle is illuminated by the lamp and the shadow produced by the lamp is illuminated by the candle. The distances between the lamp and candle and the screen are adjusted until the two shadows appear to be equally illuminated. The intensity of illumination of the screen by the candle is then equal to the intensity of illumination of the screen by the lamp. If C denotes the candle power of the candle and K that of the lamp, we have then

$$\frac{C}{r^2} = \frac{K}{r'^2},$$

where r is the distance from the candle to the screen and r' that from the lamp to the screen. If we suppose $C = 1$ we have

$$K = r'^2/r^2.$$

With this apparatus it can be shown experimentally that the intensity of illumination due to a small source varies inversely as the square of the distance. To do this first find r and r', using one candle in the way just described. Then place four candles close together, in the position previously occupied by the one candle, so that they produce only one shadow, and move the lamp nearer to the screen until the two shadows are again of equal intensity. It will be found that the distance of the lamp from the screen is half that found with one candle. If nine candles are used the lamp will have to be placed at one-third of the distance. This shows that the intensity of illumination of the screen by the lamp varies inversely as the square of the distance between them.

Another simple form of photometer is called Bunsen's grease-spot photometer. It consists simply of a piece of paper with a grease-spot on it, which can be made by melting a small piece of paraffin wax on the paper with a hot iron. The grease-spot is more transparent than the rest of the paper, so that when it is illuminated on one side the grease-spot appears brighter than the rest of the paper when viewed from the other side and less bright when viewed from the illuminated side. If both sides are equally illuminated the grease-spot is scarcely visible and both sides appear alike. To compare the candle powers of two sources they are placed one on each side of the piece of paper on a line perpendicular to its plane. The distances between the sources and the paper are then adjusted until the sides of the paper are equally illuminated, so that the grease-spot becomes almost invisible and both sides appear alike. The candle powers of the sources are then proportional to the squares of their distances from the grease-spot. This photometer is more convenient than the shadow photometer. It can also be used to prove that the intensity of illumination due to a source varies inversely as the square of the distance by comparing the source with one, four and nine candles all at the same distance from the grease-spot.

The intensity of illumination due to a source may not be the same at the same distance in all directions from the source. The candle power is then different in different directions. The average value of the candle power for all directions is called the mean spherical candle power.

REFERENCES

The Theory of Light, Preston.
The Theory of Optics, Drude.

CHAPTER II

REFLEXION AND REFRACTION AT PLANE SURFACES

WHEN light falls on a body it is usually partly absorbed by the
body and partly scattered about in all directions
Reflexion.   from the surface. Each point of the illuminated
surface scatters light in all directions, so that the surface is like a
source of light and can be seen from any direction. If the surface
on which the light falls is made of very smoothly polished silver,
very little light is absorbed and very little scattered about, but
nearly all is reflected in a definite direction. Most metals when
polished reflect light in this way, but not so completely as silver.
A clean surface of liquid mercury reflects light almost as well as
polished silver. A smooth polished surface which reflects nearly all
the light that falls on it and scatters or absorbs very little is called
a mirror.

The reflexion of light from a plane mirror can be studied with
the apparatus shown in Fig. 3, which is called an optical disk, and
can be used to illustrate clearly many of the properties of light.
The optical disk consists of a circular disk about 30 cms. in
diameter, painted white, with its circumference graduated into
degrees. The disk is mounted on a horizontal axis at its centre
about which it can be rotated and clamped in any position. PLQ
is an opaque screen also mounted independently on the same axis
as the disk. At L there is a horizontal slit about one millimetre
wide, the length of which is perpendicular to the plane of the
disk. A plane mirror MM' can be screwed to the disk so that the
axis of rotation of the disk lies along the surface of the mirror.

If a source of light is placed at S a narrow beam passes
through the slit L and falls on the mirror at the centre of the
disk. The source is placed slightly in front of the plane of the
disk so that it illuminates a narrow strip of the surface of the

disk between L and A. The beam of light is reflected by the mirror, and the reflected beam illuminates a narrow strip of the disk between A and R. The line AN is perpendicular to the mirror, and it is found that the angle NAL is always equal to the angle NAR. The angle NAL can be varied from zero to 90° by turning the disk and mirror round, keeping the screen and slit fixed. In this way we see that the incident and reflected beams of light and the normal to the mirror lie in the same

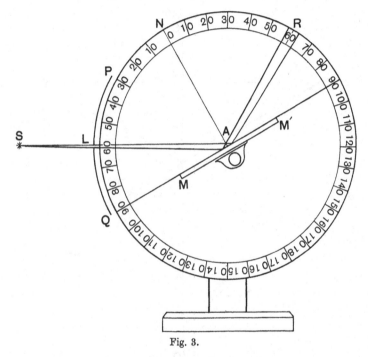

Fig. 3.

plane, and that the angle of incidence LAN is equal to the angle of reflexion NAR. If the disk is turned so that N is at L, then the mirror is perpendicular to the incident light which is then reflected back on to the slit. If the disk is then turned through an angle θ the reflected beam turns through an angle 2θ. Thus if the disk is turned so that N moves from L to the position shown in Fig. 3 we have $\theta = N\hat{A}L$, and the reflected beam AR has turned from AL through the angle LAR. But $N\hat{A}L = N\hat{A}R$,

so that $2N\hat{A}L = L\hat{A}R = 2\theta$. We see therefore that when a ray of
light is reflected from a plane mirror and the mirror is turned
through any angle the reflected ray turns through double the
angle turned through by the mirror.

If we look at a plane mirror we see an image of the objects in
front of the mirror by means of light reflected from the mirror.
The image appears to be behind the mirror. The way in which
the image is formed is shown in Fig. 4. CD represents a plane

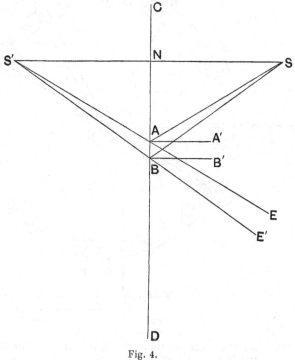

Fig. 4.

mirror, the plane of which is perpendicular to the plane of the
paper. S is a small object in front of the mirror. An eye at
EE' sees an image of S in the mirror by means of rays of light
like SAE and SBE' which are reflected from the mirror. From S
draw SN perpendicular to the surface of the mirror and produce
it to S' so that $NS' = SN$. Join ES' and $E'S'$ cutting CD at A

and B respectively. Join SA and SB. Draw AA' and BB' perpendicular to the plane CD. Then $A\hat{S}N = A\hat{S}'N$, $A\hat{S}N = A'\hat{A}S$, and $A\hat{S}'N = A'\hat{A}E$. Hence $A'\hat{A}S = A'\hat{A}E$. Therefore a ray of light travelling along SA is reflected along AE. In the same way it may be shown that the ray SB is reflected along BE'. Thus the rays reflected from the mirror into the eye seem to come from S'. The image of a point formed by a plane mirror is therefore

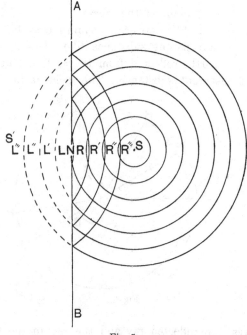

Fig. 5.

on a line through the point perpendicular to the plane of the mirror and at a distance behind the mirror equal to the distance from the point to the mirror. The rays of light reflected from the mirror seem to come from the image, but of course do not actually do so. Such an image is called a virtual image.

Instead of considering the reflexion of the rays of light from a point on the object we may consider the spherical waves which diverge from it. The rays are normal to the waves. In Fig 5

S represents a source of light and AB a plane mirror. The concentric circles round S represent successive positions of a light wave diverging from S. The dotted parts of the circles indicate where the wave would have been if the mirror had not been present. The dotted part marked L is actually in the position marked R such that $NL = NR$, for when the wave reaches the mirror it is reflected back towards S. In the same way $NR' = NL'$ and $NR'' = NL''$. The reflected wave is a sphere with centre S' at a distance NS' from the mirror equal to NS and on a line SNS' perpendicular to the plane of the mirror.

When a ray of light in air or a vacuum meets the surface of a

Refraction.

transparent substance like glass or water it is partly reflected from the surface, but part of it enters the transparent substance. The incident ray is split up

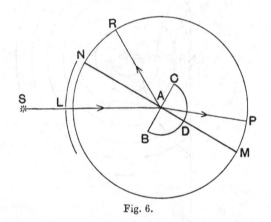

Fig. 6.

into two rays, the reflected ray and the ray in the transparent substance which is called the refracted ray. The direction of the refracted ray is not in general the same as that of the incident ray. The reflexion and refraction of a narrow beam of light at a plane glass surface can be examined with the optical disk previously described. Instead of the plane mirror a semicircular glass plate with polished sides is fastened to the disk as shown in Fig. 6. S is the source of light and L the slit as before. BCD is the glass block which should be about one cm. thick. The block is fastened to the disk so that the centre of the arc BDC is at the

centre of the disk and BC is perpendicular to the line NAM
Part of the beam SLA is reflected along AR so that $N\hat{A}L = N\hat{A}R$
as with the plane mirror.   Part of the incident beam enters the
glass at A and illuminates a narrow strip of the disk along AP.
If the disk is turned so that BC is perpendicular to LA, then the
refracted beam coincides with AM.   Thus when the incident ray
is normal to the surface of the transparent body the direction of
the refracted ray is the same as that of the incident ray.   The
beam AP is always perpendicular to the arc BDC, so that its
direction is unaltered when it passes out of the glass.

By means of the disk the relation between the angles
$L\hat{A}N$ and $P\hat{A}M$ can be examined.   It is found that $P\hat{A}M$ is
always less than $L\hat{A}N$, and that the sine of the angle LAN
bears a constant ratio to the sine of the angle PAM.   The
angle LAN is called the angle of incidence and the angle PAM
is called the angle of refraction.   Let $L\hat{A}N = i$ and $P\hat{A}M = r$,
then it is found that

$$\mu = \frac{\sin i}{\sin r},$$

where μ is a constant. μ is called the refractive index of the
glass with respect to air because the beam of light is refracted in
passing from air into glass.   For ordinary crown glass μ is equal
to about 1·52.

Whenever light passes from one medium into another across a
sharply defined boundary or surface separating the two media it is
found that $\frac{\sin i}{\sin r}$ has a constant value which is called the refractive
index of the medium containing the refracted ray with respect
to the medium containing the incident ray.   The incident and
refracted rays and the normal to the surface lie in the same plane.

If the angle of incidence is almost 90° so that the incident
ray grazes the surface of the glass, then we have $\sin i = 1$, so that

$$\mu = \frac{1}{\sin r}.$$

The angle of refraction cannot be greater than the value given by
this equation.   If $\mu = 1·52$, the greatest possible value of r is

20—2

$41° 8'$. This greatest possible value of r can easily be found with the optical disk by turning the disk so that BC is parallel to LA.

Let us now consider what happens when a ray of light travelling inside a transparent substance like glass meets the surface separating the glass from air. To examine this case with the optical disk it is only necessary to turn the disk so that the light from the slit L falls on the curved surface of the block of glass BCD as shown in Fig. 7. The beam of light SLA is perpendicular to the curved surface BDC, so that it is not deviated at this surface. At the plane surface BC it is partly reflected

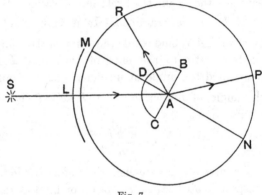

Fig. 7.

along AR and partly refracted along AP. The angle of incidence $i'' = L\hat{A}M$ is equal to the angle MAR. The angle of refraction $r' = N\hat{A}P$ is greater than the angle of incidence, and the ratio of $\sin i''$ to $\sin r'$ is found to be constant as in the previous case. The ratio $\dfrac{\sin i''}{\sin r'}$ in this case is called the refractive index of air with respect to glass, and it is found to be equal to the reciprocal of the refractive index of glass with respect to air.

If μ as before denotes the refractive index of glass with respect to air, then in the present case

$$\frac{1}{\mu} = \frac{\sin i''}{\sin r'}.$$

r' is always greater than i'', and if i'' is increased until $\mu \sin i'' = 1$, then $r' = 90°$, so that the refracted ray AP becomes parallel to

AB. If i'' is made greater than the value for which $\mu \sin i'' = 1$, the equation

$$\frac{1}{\mu} = \frac{\sin i''}{\sin r'}$$

cannot be satisfied. It is found that then there is no refracted ray and all the light is reflected at A. This is called total reflexion. If the optical disk is slowly turned round so as to increase the angle MAL, then the angle NAP slowly increases until it is equal to $90°$, and then the refracted ray disappears and the reflected ray suddenly gets brighter. The angle of incidence for which $\mu \sin i'' = 1$ is sometimes called the critical angle. If $\mu = 1\cdot52$, then the critical value of i'' is equal to $41°\ 8'$. The equation

$$\mu = \frac{\sin i}{\sin r}$$

which applies when the incident light is in the air and the equation

$$\frac{1}{\mu} = \frac{\sin i''}{\sin r'}$$

which applies when the incident light is in the glass show that the path of a ray of light through the surface of the glass is not changed when its direction of motion is reversed, for we have

$$\frac{\sin i}{\sin r} = \frac{\sin r'}{\sin i''}$$

for all possible values of i and r', hence if $i = r'$, then $r = i''$.

When a ray of light falls on a parallel plate of glass it is refracted at both surfaces of the plate and emerges parallel to its original direction. In Fig. 8 let AB be a ray of light incident on a plate of glass at B. Let i and r be the angles of incidence and refraction at B. Let the refracted ray meet the second surface at C, and let i'' and r' be the angles of incidence and refraction at C. If the sides of the plate are parallel, then $r = i''$. Also

$$\mu = \frac{\sin i}{\sin r} \quad \text{and} \quad \frac{1}{\mu} = \frac{\sin i''}{\sin r'} = \frac{\sin r}{\sin r'};$$

hence $i = r'$, so that AB and CD are parallel. If we look at an object through a thin parallel glass plate like a window, the object does not appear to be seriously distorted because the directions of the rays from the object are not altered. The sideways displacement of the rays produces some distortion, especially when the object is seen in a direction inclined to the normal to the plate.

When an object is looked at through a thick glass plate in a direction normal to the plate it appears to be nearer to the eye than if the plate were not present. In Fig. 9 let $ABCD$ be a thick glass plate with parallel sides and O an object which is viewed by an eye at E. The ray OE which is perpendicular to AB and CD passes through the plate without deviation. A ray

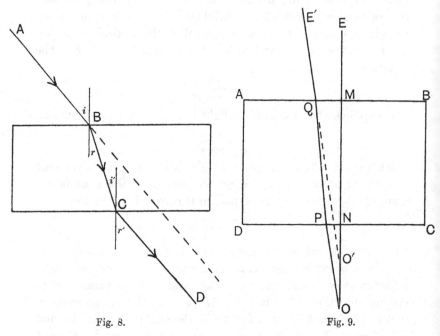

Fig. 8.   Fig. 9.

like OP is refracted at P and Q and emerges along QE' which is parallel to OP. If QE' is produced backwards it meets OE at O', so that the rays entering the eye seem to come from O' and the object O seems to be at O'. O' is the position of the virtual image of O seen by the eye. The angle PON is equal to the angle $QO'M$, and is the angle of incidence of the ray OP at P. Denote this angle by i and suppose it is very small so that $\sin i = \tan i = i$. Then we have

$$QM = iMO' = PN + rt,$$

where r is the angle of refraction at P and t is the thickness NM of the plate. Therefore since $PN = iON = i\,(OM - t)$, we get

$$iMO' = i\,(OM - t) + rt$$

or $$iOO' = it - rt.$$

But $$\mu = \frac{\sin i}{\sin r} = \frac{i}{r}$$

since i and r are supposed very small, so that

$$OO' = t\left(1 - \frac{1}{\mu}\right).$$

It appears therefore that the virtual image of the object is nearer to the eye by a distance equal to $t\left(1 - \frac{1}{\mu}\right)$. If a body at the bottom of a tank of water is viewed from above the surface of the water in a direction perpendicular to the surface, it seems to be nearer to the eye than it really is by a distance $t\left(1 - \frac{1}{\mu}\right)$ where t denotes the depth of the water. The refractive index of water with respect to air is equal to 1·33, so that the depth of the water appears to be only ¾ its real value.

The passage of a ray of light through a glass plate with plane sides which are not parallel will now be considered. Such a plate is usually called a prism. The angle between the two sides through which the ray passes is called the refracting angle of the prism. In Fig. 10 ABC represents a prism, the sides of which are supposed to be perpendicular to the plane of the paper. Let $OPQR$ be a ray of light passing through the prism in the plane of the paper, and let MPN and LQN be drawn perpendicular to AB and AC respectively. Let $O\hat{P}M = i$, $N\hat{P}Q = r$, $N\hat{Q}P = r'$, and $L\hat{Q}R = i'$. Produce OP and RQ to meet at S, and let ϕ denote the angle at S through which the ray is deviated by its passage through the prism. Let θ denote the refracting angle of the prism at A. We have

$$\phi = S\hat{P}Q + S\hat{Q}P = i - r + i' - r'.$$

Also since APN and AQN are right angles, we have

$$\theta = r + r',$$

so that $$\phi = i + i' - \theta.$$

By means of this equation and the relations

$$\mu = \frac{\sin i}{\sin r} = \frac{\sin i'}{\sin r'},$$

the value of ϕ can be computed when i, θ and μ are known. If θ is a small angle and i also small, then r, i' and r' are all small, so that approximately $\mu = i/r = i'/r'$. Hence

$$\phi = i + i' - \theta = \mu (r + r') - \theta = \theta (\mu - 1).$$

The deviation produced by a small angled prism is therefore independent of the angle of incidence on it provided this is small.

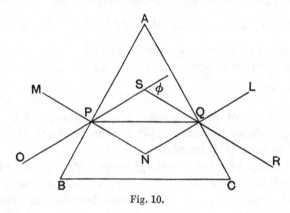

Fig. 10.

An important practical case is when PQ is equally inclined to AP and AQ, so that $r = r'$ and therefore $i' = i$. In this case $\theta = 2r$ and $\phi = 2i - \theta$, so that

$$\mu = \frac{\sin i}{\sin r} = \frac{\sin \frac{1}{2} (\phi + \theta)}{\sin \frac{1}{2}\theta}.$$

Another important case is when $i = 0$, so that $r = 0$. In this case $\theta = r'$ and $\phi = i' - \theta$, so that

$$\mu = \frac{\sin i'}{\sin r'} = \frac{\sin (\phi + \theta)}{\sin \theta}.$$

A similar case is when $r' = 0$, so that $i' = 0$. In this case $\theta = r$ and $\phi = i - \theta$, so that

$$\mu = \frac{\sin (\phi + \theta)}{\sin \theta}$$

as in the preceding case. These two cases are shown in Fig. 11. If the ray passes through the prism, then the angle r' cannot be greater than the critical angle which is about $41° 8'$ for crown glass. If C denotes the critical angle, then if r' is greater than C, the ray is totally reflected at the surface AC of the prism as

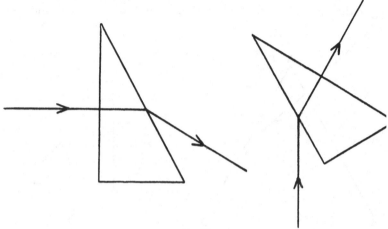

Fig. 11.

shown in Fig. 12, and so does not pass out through AC at all. If the refracting angle θ is greater than $2C$, then a ray which enters through AB for which therefore r is less than C cannot pass out through AC because $\theta = r + r'$, so that r' must be greater than C. For crown glass θ must therefore be less than $82° 16'$, or else no rays can pass through both AB and AC.

The passage of a narrow beam of light through a prism can be examined with the optical disk previously described. If the prism is fastened to the disk so that its refracting angle is close to the centre of the disk, then the deviation of the beam by the prism can be easily measured for any angle of incidence. It is found that if we start with a large angle of incidence and turn the disk slowly so as to diminish this angle, then the deviation diminishes quickly at first and then slowly and then stops diminishing and begins to increase again. The deviation is a minimum when the ray passes symmetrically through the prism, so that $i = i'$ and $r = r'$. If the minimum value of the deviation is observed, then the refractive index of the prism can be calculated by the formula

$$\mu = \frac{\sin \frac{1}{2}(\phi + \theta)}{\sin \frac{1}{2}\theta},$$

where ϕ is the minimum value of the deviation and θ the refracting angle of the prism.

A right-angled prism of crown glass is often used to reflect light through an angle equal or nearly equal to a right angle, as

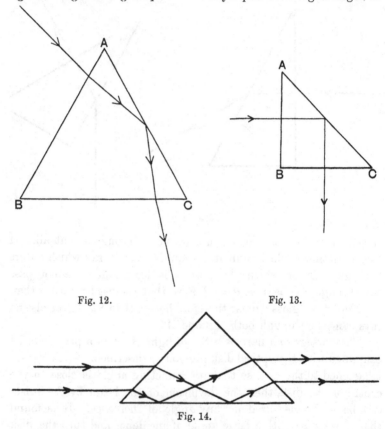

Fig. 12.   Fig. 13.

Fig. 14.

shown in Fig. 13. The ray is perpendicular to AB, so that its angle of incidence on AC is 45°, which is greater than the critical angle, so it is totally reflected through BC. Such a prism can also be used to invert rays coming from an object as shown in Fig. 14. The rays are totally reflected on the base and emerge parallel to their original directions.

CHAPTER III

SPHERICAL MIRRORS

A SPHERICAL mirror is a polished silver or other reflecting surface in the form of a portion of the surface of a sphere. Spherical mirrors are often made by grinding one surface of a circular disk of glass to a spherical form, polishing it and then silvering it. Fig. 15 shows cross sections of a concave and a convex mirror. The line from the middle point of a circular spherical mirror to its centre of curvature is called the axis of the mirror. In Fig. 16 let AB be a concave mirror, O its middle point and R its centre of

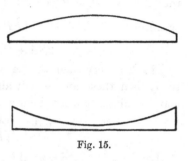

Fig. 15.

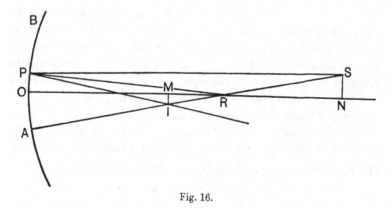

Fig. 16.

curvature. Let S be a small source of light at a short distance SN from the axis ORN of the mirror. Join SR and produce it to meet the mirror at A. A ray of light from S travelling along SA will be reflected back along AS because RA is perpendicular to the surface of the mirror at A. Now take any point P on the mirror and join RP. A ray SP will be reflected at P so that the reflected ray is in the same plane as SP and RP and makes an angle with RP equal to the angle SPR. Let PI be this reflected ray. The two rays from S reflected at A and P respectively meet at I.

We have $\qquad\qquad P\hat{I}A = I\hat{P}R + P\hat{R}A$

and $\qquad\qquad\qquad P\hat{R}A = R\hat{P}S + P\hat{S}A$;

therefore $\quad P\hat{S}A + P\hat{I}A = I\hat{P}R + P\hat{R}A + P\hat{R}A - R\hat{P}S,$

or since $\qquad\qquad\qquad I\hat{P}R = R\hat{P}S,$

$$P\hat{S}A + P\hat{I}A = 2P\hat{R}A.$$

If SN is very small so that OA is small, and if P is very near to A, then these angles will all be small so that their tangents may be substituted for them without serious error. Hence

$$\frac{PA}{AS} + \frac{PA}{AI} = \frac{2PA}{AR}.$$

Let $\qquad\qquad AS = u,\ AI = v,\ \text{and}\ AR = r,$

then $\qquad\qquad\qquad \frac{1}{u} + \frac{1}{v} = \frac{2}{r}.$

Since SN is supposed very small, u and v are equal to the distances of S and I from the centre of the mirror O. This shows that the position of I is independent of the position of P so that all rays from S which fall on the mirror near to A and O will be reflected so that they pass through I. The rays are said to come to a focus at I and there is said to be a real image of S at I.

Since the distance of I from O is independent of SN, so long as SN is small, it follows that the image of N is at M and MI is the image of the line NS, for each point on NS has an image on MI. The ray SO is reflected along OI and $S\hat{O}N = M\hat{O}I$, hence

$$\frac{SN}{IM} = -\frac{ON}{OM} = -\frac{u}{v}.$$

The minus sign is required because NS is upwards and MI downwards. The height of the image is therefore to the height of the source in the ratio of the distance of the image from the mirror to the distance of the source from the mirror. The image it will be observed is inverted. The equation

$$\frac{1}{u} + \frac{1}{v} = \frac{2}{r}$$

gives the distance of the image from the mirror corresponding to any position of the source so long as the source is near the axis of the mirror. If $u = r$ we get $v = r$ so that $u = v$. Hence an object near the axis at a distance from the mirror equal to the radius of curvature of the mirror gives an image at the same distance from the mirror. This image is inverted and equal to the object. This is shown in Fig. 17. The image of N is at N but the image of S is at I, and $NS = NI$.

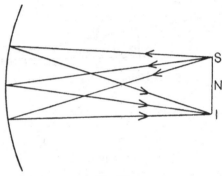

Fig. 17.

If the source is nearer to the mirror than the centre of curvature so that $u < r$, we get $v > r$. In this case the image is further from the mirror than the source. If $u = r/2$ we get $1/v = 0$, so that v becomes indefinitely large. This means that the reflected rays are parallel so that they never come to a focus. This case is shown in Fig. 18.

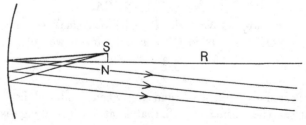

Fig. 18.

If the source is at a very great distance from the mirror we have $1/u = 0$ so that $v = r/2$. In this case the rays falling on the mirror are parallel and are brought to a focus at a distance from the mirror equal to $r/2$. One-half the radius of curvature is often called the focal length of the mirror. If u is less than $r/2$, then the formula

$$\frac{1}{u} + \frac{1}{v} = \frac{2}{r}$$

makes v negative. This indicates that there is no real image but that the rays after reflexion seem to come from a point behind the mirror at a distance from it equal to v. There is said to be a

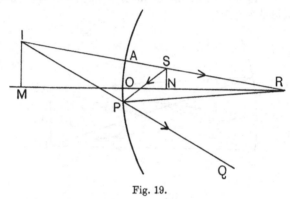

Fig. 19.

virtual image at this point. This case is shown in Fig. 19. We have

$$R\hat{P}Q = P\hat{I}A + P\hat{R}A,$$

and

$$P\hat{S}A = P\hat{R}A + S\hat{P}R.$$

Also

$$S\hat{P}R = R\hat{P}Q.$$

Hence

$$P\hat{S}A - P\hat{I}A = 2P\hat{R}A.$$

As before we suppose NS and OP to be very small, so that if we put AI or $OM = -v$, AS or $ON = u$ and $OR = r$, we get

$$\frac{1}{u} + \frac{1}{v} = \frac{2}{r}.$$

The reflected rays like AR and PQ when produced backwards meet at the virtual image of S, that is at I. The image is erect and larger than the object.

Let us now consider the reflexion of light from a convex mirror. This case is shown in Fig. 20. R is the centre of curvature and RON the axis of the mirror. Let S be a small source of light. Join SR cutting the mirror at A. The ray SA is reflected back

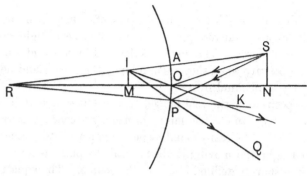

Fig. 20.

along AS. Take any point P near to A and O. The ray SP is reflected along PQ; QP when produced meets RA at I.

We have $\qquad P\hat{S}A + P\hat{R}A = S\hat{P}K,$

$$P\hat{I}A = P\hat{R}A + I\hat{P}R.$$

Also $\qquad S\hat{P}K = K\hat{P}Q = I\hat{P}R.$

Therefore $\qquad P\hat{S}A - P\hat{I}A = -2P\hat{R}A.$

Let $\qquad OR = -r,\ AI = -v$ and $AS = u,$

so that as before when SN is very small and P is very near to O, we get

$$\frac{AP}{u} - \frac{AP}{-v} = -2\frac{AP}{-r},$$

or $\qquad \dfrac{1}{u} + \dfrac{1}{v} = \dfrac{2}{r}.$

The convex mirror produces a virtual image which is smaller than the object and is erect. We have

$$\frac{MI}{NS} = \frac{OM}{ON} = -\frac{v}{u}.$$

It appears that the equations

$$\frac{1}{u} + \frac{1}{v} = \frac{2}{r}$$

and
$$\frac{NS}{MI} = -\frac{u}{v}$$

apply without modification to all cases of reflexion from either concave or convex mirrors provided that only rays of light near the axis are considered. The distances denoted by u, v and r must be reckoned positive when they are in front of the mirror and negative when they are behind it.

The position of the real image, of a bright source of light like an electric lamp, formed by a concave mirror can be easily found by putting up a white screen in the reflected light and moving it about until the place is found where there is a sharply defined image of the source. The equations

Experiments with Spherical Mirrors.

$$\frac{1}{u} + \frac{1}{v} = \frac{2}{r}$$

and
$$\frac{NS}{MI} = -\frac{u}{v}$$

can be verified by measuring u and r and the size of the image with the source at different distances from the mirror.

A plane white screen with a hole at its centre covered with wire gauze may be used to find the radius of curvature of a concave mirror. The screen is put up in front of the mirror so that it is perpendicular to the axis and the hole is close to the axis. The hole is illuminated from behind by an electric lamp and the screen is moved along the axis of the mirror until a sharp image of the gauze is formed on the screen close to the hole in it. The distance from a point half way between the image and the hole to the mirror is then equal to the radius of the mirror. The radius of curvature of a convex mirror is best found by methods involving the use of lenses which will be described in the following chapter.

So far we have considered only incident and reflected rays of light making very small angles with the axis of the mirror and near to it, and have calculated the size and position of the image formed by such rays. It is found that

Focal Lines.

rays making large angles with the axis do not all pass through the same point after reflexion, so that a definite image is not produced by a spherical mirror when such rays are employed.

In Fig. 21 let MCM' be a spherical mirror with centre of curvature at O and S a small source of light. Let SMF_1F_2 and $SM'F_1F_2'$ be rays from S. Join OM and OM' and produce SO cutting the reflected rays at F_2 and F_2'. Since SO is part of a diameter of the spherical surface it is easy to see that all rays from S will cut SO after reflexion. All the reflected rays therefore cut the line F_2F_2'. If a white screen is put up at F_2F_2' we get a line of light on it. This line is called the second focal line. Also if

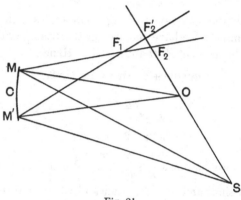

Fig. 21.

we imagine the plane $SM'MF_2$ to be rotated about OS as axis we see that the distance of F_1 from OS will not be altered so that all the reflected rays pass through a line at F_1 perpendicular to the plane of the paper. This line is called the first focal line. The distances of these focal lines from the mirror can be easily calculated. Let

$$CO = r, \ CS = u, \ CF_1 = v_1, \ CF_2 = v_2.$$

Also let the angle of incidence $OCS = i$.

We have $M\hat{O}M' + O\hat{M}S = O\hat{M}'S + M\hat{S}M'$,

and $M\hat{O}M' + O\hat{M}'F_1 = O\hat{M}F_1 + MF_1M'$;

also $O\hat{M}S = O\hat{M}F_1$

and $O\hat{M}'S = O\hat{M}'F_1.$

Adding up these equations we get

$$2\widehat{MOM}' = \widehat{MSM}' + \widehat{MF_1M}'.$$

If we suppose that the diameter MM' of the mirror is small, then these angles are small so that they may be replaced by their tangents. Now

$$\tan MOM' = MM'/r, \quad \tan MSM' = MM' \cos i/u,$$

and
$$\tan MF_1M' = MM' \cos i/v_1;$$

hence
$$\frac{1}{u} + \frac{1}{v_1} = \frac{2}{r \cos i}.$$

If $i = 0$ this becomes $\dfrac{1}{u} + \dfrac{1}{v_1} = \dfrac{2}{r}$ in agreement with the result previously obtained for rays making small angles with the axis.

To determine v_2 we have the area of the triangle SCF_2 equal to the sum of the areas of SCO and OCF_2. Hence

$$ur \sin i + v_2 r \sin i = uv_2 \sin 2i,$$

which gives
$$\frac{1}{u} + \frac{1}{v_2} = \frac{2 \cos i}{r},$$

since
$$\sin 2i = 2 \cos i \sin i.$$

If $i = 0$ this also becomes $\dfrac{1}{u} + \dfrac{1}{v_2} = \dfrac{2}{r}$. When i is small the two lines are very short and near together and when $i = 0$ they coincide and form a single point, the image of S. The way in which the reflected rays pass through the two focal lines when the angle of incidence is not very large is shown approximately in Fig. 22. AB

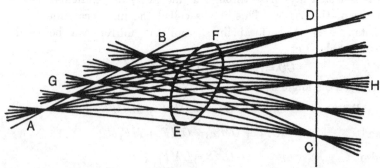

Fig. 22.

is supposed to represent a horizontal line and CD a vertical line. Both these lines are perpendicular to GH which cuts them at their middle points. Five rays are shown passing through each of five equidistant points on AB. The five rays from each point on AB go to five equidistant points on CD and the five rays passing through each point on CD go to the five points on AB. All the rays pass through a circle EF the centre of which is on GH and lies in a plane perpendicular to GH. This circle is about half-way between the focal lines. If a screen is put up in the plane of this circle, a circular patch on it is illuminated. This patch is the nearest approach to an image of S that the reflected rays form. Unless the patch is very small the mirror does not form a distinct image of an object at S.

CHAPTER IV

LENSES

Refraction at a
Spherical
Surface.
THE following construction enables the refracted ray to be
drawn when a ray of light in air meets a transparent sphere. Let AD (Fig. 23) be a sphere of
glass or other transparent substance, the refractive index of which
with respect to air is μ. Let PA be a ray of light in air meeting

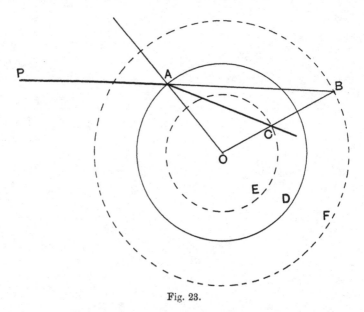

Fig. 23.

the sphere at A. Let the radius of the glass sphere be r and let its
centre be at O. With centre O describe two spheres EC and FB
having radii equal to r/μ and μr respectively. Produce PA to
meet the sphere of radius μr at B. Join OB cutting the sphere of

radius r/μ at C. Join AC. The ray PA is refracted at A and travels along AC. We have

$$\frac{OA}{OC} = \mu \ \text{ and } \ \frac{OB}{OA} = \mu,$$

so that the triangles OAC and OBA are similar triangles and therefore the angle CAO is equal to the angle ABO. But

$$\frac{\sin BAO}{\sin ABO} = \frac{OB}{OA} = \mu,$$

and therefore $$\mu = \frac{\sin BAO}{\sin CAO},$$

which shows that AC is the refracted ray corresponding to PA. This construction shows that all rays in the air like PA which when produced pass through the point B are refracted at the surface of the glass sphere so that they pass through the point C. If there were a small source of light at C inside the sphere, then a ray from C like CA would be refracted at the surface of the sphere and would travel along AP. Thus all rays from C like CA after refraction appear to come from B so that B is the virtual image of C formed by refraction at the spherical surface.

A lens is a circular disk of glass the surfaces of which are spherical in shape. The centres of curvature of
Lenses. the two surfaces and the middle point of the lens should be in a straight line which is called the axis of the lens. A lens forms an image of an object near to its axis which may be real or virtual as with spherical mirrors. A thin lens is one in which the distance between the spherical surfaces at the axis is very small compared with their radii of curvature. The position of the image formed by a thin lens can be easily calculated. In Fig. 24 let AB be a thin lens, O and O' the centres of curvature

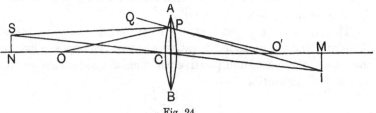

Fig. 24.

of its surfaces and OCO' its axis. Let S be a small source of light
very near to the axis. The ray SCI meeting the lens at its centre
is not deviated by the lens because at the centre the two surfaces
of the lens are parallel, and a thin parallel plate does not deviate
a ray of light, as we have seen. Take any point P in the lens near
to C. The ray SP is deviated by the lens and meets the ray SCI
at I. Produce IP to Q. The angle QPS is equal to the angle
through which the ray SPI is deviated by the lens. At P the
angle between the two surfaces of the lens is equal to the sum of
the angles POC and $PO'C$ because OP and $O'P$ are perpendicular
to the surfaces at P. When a ray passes through a small-angled
prism the angle between the two surfaces of which is θ, then the
ray is turned through an angle equal to $(\mu - 1)\,\theta$ where μ is the
refractive index of the prism. Hence the angle QPS is equal to

$$(\mu - 1)\ (P\hat{O}C + P\hat{O'}C),$$

where μ is the refractive index of the lens with respect to air.
But the angle QPS is equal to

$$P\hat{S}C + P\hat{I}C,$$

so that $\qquad P\hat{S}C + P\hat{I}C = (\mu - 1)(P\hat{O}C + P\hat{O'}C).$

Since PC and SN are very small we can replace these angles by
their tangents without serious error. Let

$$CS = u, \ -CI = v, \ CO = r \ \text{and} \ -CO' = r'.$$

Then we get

$$\frac{CP}{u} + \frac{CP}{-v} = (\mu - 1)\left(\frac{CP}{r} + \frac{CP}{-r'}\right),$$

or $\qquad\qquad \dfrac{1}{u} - \dfrac{1}{v} = (\mu - 1)\left(\dfrac{1}{r} - \dfrac{1}{r'}\right).$

This equation shows that v is independent of the position of P, so
long as PC and SN are small, so that all rays from S pass through
I, which is therefore the image of S.

The lengths u, v, r and r' are all measured from the centre of
the lens C and are reckoned negative when on the opposite side of
the lens to the source and positive when on the same side as the
source. The quantity

$$(\mu - 1)\left(\frac{1}{r} - \frac{1}{r'}\right)$$

is a constant and is equal to the reciprocal of what is called the focal length of the lens. If the focal length is f, then

$$\frac{1}{f} = (\mu - 1)\left(\frac{1}{r} - \frac{1}{r'}\right)$$

and

$$\frac{1}{u} - \frac{1}{v} = \frac{1}{f}.$$

If u is very large, then $-v$ is equal to the focal length. For example if an object and its image are on opposite sides of a lens and both 50 cms. from it, then

$$u = +50 \text{ and } v = -50,$$

so that

$$\frac{1}{50} + \frac{1}{50} = \frac{1}{f},$$

which gives

$$f = +25 \text{ cms.}$$

A lens which diminishes the divergence of rays is called a convex lens and can form a real image. The focal length of such a lens is reckoned to be positive. A lens which causes rays to diverge more is called a concave lens and has a negative focal length. A concave lens can only form a virtual image like a concave mirror. The equation

$$\frac{1}{u} - \frac{1}{v} = (\mu - 1)\left(\frac{1}{r} - \frac{1}{r'}\right)$$

applies to any thin lens in any case so long as the object is near to the axis of the lens and only rays near the axis are considered. With a convex lens, if $u = f$, we get

$$\frac{1}{f} - \frac{1}{v} = \frac{1}{f},$$

so that v is indefinitely large. This means that the rays after passing through the lens are parallel and so never converge to a focus. If u is less than f, then v is positive, which means that the rays still diverge after passing through the lens. They seem to come from a virtual image of the object on the same side of the lens as the object. This case is shown in Fig. 25. For example if $u = 5$, $f = 7$, we get

$$\frac{1}{5} - \frac{1}{v} = \frac{1}{7},$$

which gives $v = +17\cdot5$, so that the image is on the same side of the lens as the source and the image is therefore a virtual one. If $u = 7$, $f = 5$, we get

$$\frac{1}{7} - \frac{1}{v} = \frac{1}{5},$$

which gives $v = -17\cdot5$, which shows that the image is on the opposite side of the lens to the object and so is a real image.

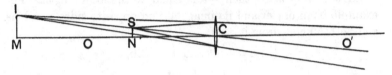

Fig. 25.

A concave lens can give only a virtual image. For example if $u = 10$ and $f = -10$ we get

$$\frac{1}{10} - \frac{1}{v} = -\frac{1}{10},$$

so that $v = +5$. The image is therefore on the same side as the object and is a virtual one. This case is shown in Fig. 26. In all cases it is easy to see that

$$\frac{NS}{MI} = \frac{u}{v}.$$

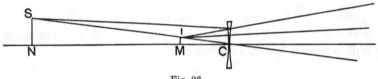

Fig. 26.

To find graphically the position of the image of a point formed by a lens of given focal length, the simplest way is to draw a ray from the point through the centre of the lens and another ray

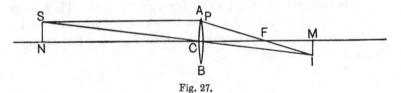

Fig. 27.

parallel to the axis. In Fig. 27 let AB be a thin lens with centre C and let NC be its axis. Let S be the point of which the image is required. Draw SC and produce it. Draw SP parallel to NC. From NC, produced if necessary, cut off CF equal to the focal length, so that F is on the opposite side of the lens to S if the focal length is positive and on the same side if negative. Join PF and produce it, if necessary, to meet SC at I, which is the image of S. Fig. 28 shows the same thing when the focal length is negative.

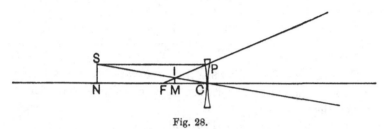

Fig. 28.

For a crown glass lens $\mu = 1\cdot52$ about, so that

$$\frac{1}{f} = 0\cdot52 \left(\frac{1}{r} - \frac{1}{r'} \right).$$

If both sides of the lens are convex and of equal curvature then

$$r' = -r,$$

so that
$$\frac{1}{f} = 0\cdot52 \frac{2}{r} \quad \text{or} \quad f = \frac{r}{1\cdot04}.$$

The focal length of a thin convex lens can be found by using a bright source and forming a real image of it on a white screen by means of the lens. As source a piece of wire gauze strongly illuminated from behind by an electric lamp may be used. The gauze is put up perpendicular to the axis of the lens at a distance from the lens not less than f. A white screen is put up on the other side of the lens and moved along until the position is found at which there is a sharp image of the gauze on it. If the distance from the gauze to the middle of the lens is 30 cms. and from the lens to the screen 50 cms., then

$$u = 30 \quad \text{and} \quad v = -50,$$

so that
$$\frac{1}{30} + \frac{1}{50} = \frac{1}{f},$$

which gives
$$f = 18\cdot75 \text{ cms.}$$

If the distance between the gauze and screen is greater than $4f$, then there are two positions of the lens between them at which it gives a sharp image of the gauze on the screen. A good way of finding f is to measure the distance d between the gauze and the screen and also the distance c between these two positions of the lens.

Then it is easy to see that

$$u = \frac{d-c}{2} \text{ or } \frac{d+c}{2},$$

and

$$-v = \frac{d+c}{2} \text{ or } \frac{d-c}{2}.$$

For example if the distance from the gauze to the screen is 120 cms. and the distance between the two positions of the lens is 20 cms., then

$$u = \frac{120-20}{2} = 50,$$

and

$$-v = \frac{120+20}{2} = 70,$$

which give

$$f = 29{\cdot}17 \text{ cms.};$$

or we may take

$$u = \frac{120+20}{2} = 70,$$

and

$$-v = \frac{120-20}{2} = 50,$$

which give

$$f = 29{\cdot}17 \text{ cms.}$$

as before. If the distance between the gauze and the screen is made gradually smaller, then c becomes smaller until there is only one position of the lens at which it gives a distinct image on the screen. The lens is then half-way between the gauze and screen so that $u = -v$, and therefore

$$\frac{1}{u} + \frac{1}{u} = \frac{1}{f} \text{ or } u = 2f.$$

Hence in this case $d = 4f$. If d is made less than $4f$, no distinct image can be obtained on the screen.

The radius of curvature of a convex mirror can be found by means of a convex lens. The lens is arranged so that it produces a sharp image of the gauze on the screen. The convex mirror is then put up between the lens and screen so that it reflects the light back through the lens. The mirror is moved along until a sharp image of the gauze is produced at the same distance from the lens as the gauze. The gauze for this experiment should cover a small hole in a white screen so that the image can be seen on this screen close to the gauze. The distance between the convex mirror and the first screen on which the lens forms a sharp image of the gauze when the mirror is removed is then equal to the radius of the mirror. This experiment is shown in Fig. 29.

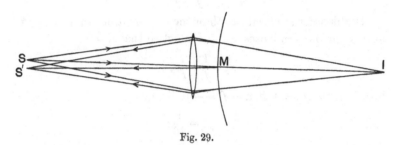

Fig. 29.

S' is the image of S formed by reflexion from the mirror back through the lens. I is the image of S formed by the lens when the mirror is removed. Obviously IM is equal to the radius of curvature of the mirror, for the rays are perpendicular to its surface.

The focal length of a concave lens can be found by combining it with a convex lens of shorter focal length and measuring the focal length of the combination, which acts as a convex lens.

Suppose that two lenses with focal lengths f and f' are placed so that their axes are in the same straight line and their centres at a distance d apart. Let there be a source on the axis at a distance u from the lens nearest to it, the focal length of which is f. This lens alone would form an image of the source at a distance v from it such that

$$\frac{1}{u} - \frac{1}{v} = \frac{1}{f}.$$

The light falling on the second lens comes from this image which is at a distance $v + d$ from it. Hence the second lens forms an image at a distance v' from itself given by the equation

$$\frac{1}{v+d} - \frac{1}{v'} = \frac{1}{f'}.$$

But $v = uf/(f-u),$

so that $\dfrac{1}{\dfrac{uf}{f-u}+d} - \dfrac{1}{v'} = \dfrac{1}{f'}.$

If $d = 0$ this reduces to

$$\frac{1}{u} - \frac{1}{v'} = \frac{1}{f} + \frac{1}{f'}.$$

If F denotes the focal length of the combination we have therefore, when the two lenses are in contact so that $d = 0$,

$$\frac{1}{F} = \frac{1}{f} + \frac{1}{f'}.$$

For example if $f = +20$ and $f' = -40$ we get

$$\frac{1}{F} = \frac{1}{20} - \frac{1}{40},$$

which gives $F = +40$, so that the combination acts like a convex lens with focal length 40. If $f = -f'$, then F becomes indefinitely large, which shows that the combination acts like a piece of plane glass.

The equation

$$\frac{1}{f} = (\mu - 1)\left(\frac{1}{r} - \frac{1}{r'}\right)$$

enables the refractive index of the lens to be calculated when f, r, and r' are known. The radii r and r' can be measured with a spherometer and f found as above described. A simple way of finding the radii when a spherometer is not available is the following, shown in Fig. 30. The lens is made to float on a small quantity of mercury contained in a shallow glass vessel. It may be fixed in position with some wax if necessary. This converts the lower surface of the lens into a mirror. A pin NS is put up above the lens so that its point S lies on the axis of the lens. The pin is moved up and down until the image of the point of the

pin formed by reflexion at the lower surface of the lens coincides
with the point itself. The image
of the pin is then at SM. When
the pin and image coincide their
apparent relative positions do not
change when the eye looking at
them is moved about. The rays
from S return to S after being
reflected, so that their angle of
incidence on the lower surface of
the lens must be zero. The rays
in the lens, if produced, therefore
meet at the centre of curvature O
of the lower surface. Let this radius
be denoted by r and the distance
from S to the centre of the lens by
u, then we have

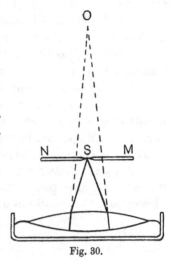

Fig. 30.

$$\frac{1}{u} - \frac{1}{r} = \frac{1}{f},$$

where f is the focal length of the lens. O is the virtual image
of S formed by the lens because if the mercury were removed the
rays passing through the lens would seem to come from O.

For example if $u = 15$ and $f = 30$ we get

$$\frac{1}{15} - \frac{1}{r} = \frac{1}{30},$$

so that $r = 30$. The other radius r' can be found in the same way
by turning the lens over. Suppose $r' = -30$, then we have

$$\frac{1}{30} = (\mu - 1)\left(\frac{1}{30} + \frac{1}{30}\right),$$

which gives $\mu = 1\cdot50$.

So far we have considered only rays of light very near to the
axis of the lens and have calculated the position of the image
formed by such rays. We have also supposed the lens to be very
thin.

In Fig. 31 let LMN be a large lens having one side plane and
the other convex. Let R_3, R_2, R_1, R, R', R_1', R_2', R_3' be parallel

rays of light falling on the plane side of the lens and perpendicular to it. The rays R and R' which are near to the axis are brought to a focus at F. The rays R_1 and R_1' meet at F_1, which is further from the lens than F. The rays R_2 and R_2' meet at F_2, and R_3 and R_3' at F_3. Such a lens does not form a distinct image of an object because all the rays from a point on the object do not meet at a single point after passing through the lens. The rays further from the axis meet further from the lens than those nearer the axis. A distinct image can be obtained by covering the lens with a screen having a small aperture in it at the centre of the lens, so that only rays close to the axis are able to get through.

The distance between the point F at which parallel rays near the axis meet after passing through the lens and a point like F_3 at which parallel rays a distance r from the axis meet is called the

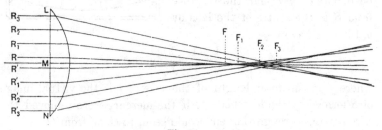

Fig. 31.

spherical aberration of the lens at a distance r from its axis. Lenses intended to form distinct images have to be designed so as to diminish the spherical aberration to a very small amount. This can be done by giving the surfaces suitable curvatures and if necessary using a combination of several lenses instead of one. The spherical aberration of a lens is less when it is arranged so that both surfaces deviate any ray equally. For example the lens shown in Fig. 31 has much less spherical aberration if its curved surface is turned towards the parallel rays.

When the light from a small source passes through a lens in a direction inclined to the axis of the lens, then the rays pass through two perpendicular focal lines as with a concave mirror. These lines are shorter and nearer together the smaller the inclination to the axis.

CHAPTER V

DISPERSION

WHEN examining the deviation of a narrow beam of white light by a prism with the optical disk it may be noticed that the beam after passing through the prism shows signs of colour. The beam appears red on the side which is least deviated and blue on the other side. This phenomenon can be studied with the apparatus shown in Fig. 32. S is a powerful source of white light like

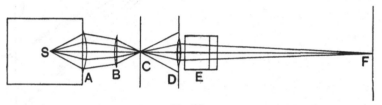

Fig. 32.

an arc lamp, contained in a box. A is a large convex lens which serves to concentrate light from S on to a second lens B by which the light is focused on a screen C which contains a narrow vertical slit. The slit is thus very strongly illuminated. The light from the slit falls on a screen D at the centre of which is a convex lens. This convex lens is placed so that the light from the slit at C is focused on a white screen F. In this way an intensely illuminated image of the slit is focused on the screen at F. If now a dense flint glass prism E is put in front of the lens D so that the light passes through the prism, the light is refracted by the prism and the image of the slit appears in a new position. The edge of the prism at its refracting angle should be parallel to the slit so that the light is refracted in a plane perpendicular to the length of the slit. A plan of the same apparatus is shown in Fig. 33. When the prism is put up the image of the slit C moves from I to PQ and is drawn out in a direction perpendicular to the length of the slit into a band of different colours which is called the spectrum of the white light. The colours of the band in order starting at P are red, orange, yellow, green, blue and violet. As we pass along the spectrum from P to Q the colour gradually changes in quality so that the red changes into orange and yellow without any sudden variation, and the yellow changes gradually into greenish yellow and then green and so on. Thus there is an indefinitely large number of shades of colour in the spectrum. A

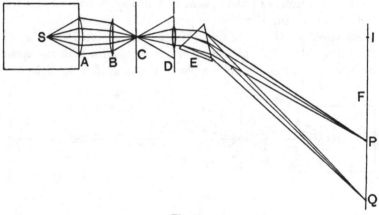

Fig. 33.

similar experiment was tried by Newton, who supposed that ordinary white light is a mixture of different coloured lights which are separated from each other by the prism because lights of different colours are deviated through different angles. The red is least deviated and the violet most. This shows that the refractive index of the prism with respect to air is greater for violet light than for red light. The formula

$$\mu = \frac{\sin \frac{1}{2} (\phi + \theta)}{\sin \frac{1}{2} \theta}$$

shows that the minimum deviation ϕ is greater when μ is greater. The spreading out of white light into a spectrum by refraction is called dispersion.

If the different coloured lights which make up the spectrum are mixed up again they produce white light. This can be shown by putting a convex lens between the prism and the screen in such a position that it throws an image of the face of the prism from which the light issues upon the screen. All the different coloured lights are then concentrated on to this image which appears white. Another way is to place a large concave mirror at *F* instead of the screen and let the light reflected from it fall on a suitably placed screen. It is found that the mirror and screen can be arranged so that the whole spectrum is concentrated into a patch of white light, which is an image of the face of the prism *E*.

If a narrow slit is made in the screen *F* it can be arranged so that light of any desired colour alone passes through it. Suppose the slit is put so that it lies across the yellow part of the spectrum, then we get a beam of yellow light on the other side of the screen. If this beam of yellow light is focused on another screen with a convex lens we get a yellow image of the slit in *F*. If a prism is put up in front of the lens the beam of yellow light is deviated by the prism but it is not drawn out into a spectrum. The yellow image of the slit is moved to a new position but its appearance is unchanged. By moving the slit in *F* along the spectrum this experiment can be tried with all the different colours. It is found that the result is the same with all of them. The prism deviates the violet much more than the red but it does not split either up into new colours.

It is found that the wave lengths of the different coloured lights in the spectrum of white light are different. As we pass along the spectrum from violet to red the wave length increases. The following table gives the wave lengths of the different colours in millionths of a millimetre.

Colour			Range of Wave Length in Air
Red	..	...	647–810
Orange	...	...	586–647
Yellow	...	...	535–586
Green	...	...	492–535
Blue	...	...	424–492
Violet	...	...	360–424

The methods by which these wave lengths have been determined will be described in a later chapter. The light at any
point in a spectrum consists of trains of waves in the ether having
a definite wave length (λ) and frequency (n). We have $v = n\lambda$
where v denotes the velocity of light. The velocity v is the same
in a vacuum for light of any wave length. In glass and other
transparent substances the velocity depends on the wave length.
The velocity usually but by no means always is smaller the
smaller the wave length. When a train of light waves passes
from one medium into another in which its velocity is different
the wave length is changed but the frequency remains the same.
If the velocity and wave length in the first medium are denoted
by v and λ and in the second medium by v' and λ', then we have

$$v = n\lambda \quad \text{and} \quad v' = n\lambda',$$

so that
$$\frac{v}{\lambda} = \frac{v'}{\lambda'},$$

where n denotes the frequency or number of vibrations per second
at any fixed point while the train of waves is passing over it. It
is found that the colour of light does not depend on its intensity.
According to the wave theory the intensity of light depends on
the amplitude of the vibration in the ether and the colour on the
wave length or frequency.

The only difference between violet light and red light is that
the wave length of violet light is about half that of red light.
The light emitted by very hot solid bodies gives a spectrum like
that just described which contains an indefinitely large number of
differently coloured lights and so forms a continuous band of
colour from the red end to the violet. If a slit is illuminated
with light of only one wave length, then the spectrum of this light
obtained with a lens and prism consists of a single image of the
slit. The many coloured spectrum of white light from a hot solid
body consists of a series of images of the slit infinite in number,
one for each shade of colour. Each image of the slit has a certain
width so that at any point in the spectrum an indefinitely large
number of these images overlap. At any point in the spectrum
there is therefore not light of one wave length but light of many
wave lengths which however all fall between limits near together.

The narrower the slit the fewer images of it overlap at any point. The spectrum of white light is called a continuous spectrum and when a very narrow slit is used it is said to be purer than when a wide slit is used, because then the range of wave lengths at any point is smaller. A pure spectrum is a spectrum in which at any point there is only light of one wave length present. An approximation to a pure spectrum can be obtained by using a very narrow slit and focusing the spectrum accurately on the screen so that the images of the slit are sharp and narrow.

When the spectrum of the light emitted by gases is examined it is found that it does not consist of a continuous spectrum but of a number of separate images of the slit. Such a spectrum is called a line spectrum because when the slit is narrow each image looks like a line. If we use a mercury arc lamp as the source of light S in the apparatus shown in Fig. 32 and Fig. 33, the spectrum obtained when the prism is put up consists of a number of coloured images of the slit of which the four brightest are red, green, blue and violet respectively. These images occupy the same positions on the screen as the same colours did when white light was used and a continuous spectrum obtained. The light from a mercury arc lamp is emitted by mercury vapour through which a current of electricity is passing. The line spectrum of the light emitted is called the spectrum of mercury. Each line in it consists of light of a definite wave length. It is found that the vapours of other elements when emitting light also give line spectra. For example, sodium vapour in a flame gives two bright yellow lines very close together, thallium vapour gives a single bright green line, and lithium a bright red line. Some elements, for example iron, give spectra containing many hundreds of lines. By examining the spectrum of a source of light giving a line spectrum the elements present in the source can be determined by comparing the lines observed with those known to be produced by different elements. This method of finding the composition of a body is called spectrum analysis. Instruments for observing spectra are called spectroscopes and spectrometers. A simple form of spectrometer is shown in Fig. 34. LC is a brass tube having a convex lens L at one end and a slit C at the other. The end of the tube carrying the slit can slide in and out so that the distance from the slit to the lens can be varied. The length of the slit is perpendicular to the plane of the paper. This tube is called a collimator. The collimator is fixed to a circular table T on which a prism P can be placed. MN is another brass tube having a convex lens at M and a small convex lens at N fixed to a tube which can slide in and out. This tube with its lenses is called a telescope. The telescope is carried on an arm which can be turned about an axis at the centre of the table T and perpendicular to its plane. The circumference of the table is graduated into degrees and fractions

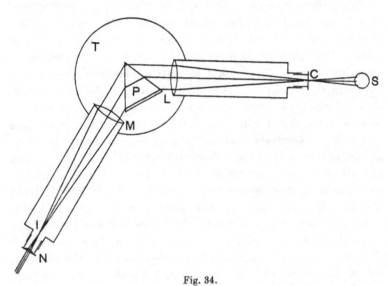

Fig. 34.

of a degree, and the arm carrying the telescope has a vernier attached to it by means of which the angular position of the telescope on the graduated circle can be determined. The prism P is attached to a circular plate which can also be turned about an axis at the centre of the table perpendicular to the plane of the table. This plate has a vernier attached to it by means of which the angular position of the prism can be found. If a source of light such as a flame or a mercury arc lamp is put up at S in front of the slit C, the rays of light which pass through each point of the slit and fall on the lens L should be made parallel by this lens. The distance of the slit from the lens L has to be adjusted so that it is equal to the focal length of the lens. The beams of parallel rays pass through the prism as shown. If the light from any one point on the slit is all of the same colour it is all deviated by the prism to the same extent and so remains a parallel beam after passing through the prism. If several different colours are present then they are deviated through different angles, so that we get as many separate parallel beams as there are colours or wave lengths present in the light. If the telescope is turned so that the axis of the lens M is parallel to one of these parallel beams, then the parallel rays pass through the lens M and are brought to a focus at I inside the telescope. At I we get a real image of the slit C. This image can be observed through the lens N which magnifies it. At I a fine cross wire is usually stretched across the tube so that it coincides with the image I when this is in the middle of the tube. The angular position of the telescope when it is turned so that the wire coincides with an image of the slit can be read off. If the spectrum of the light from the source S contains many lines then several of them may be visible in the telescope at once and by turning the telescope round they can all be seen in turn. The deviation of the light by the prism depends on the angle of incidence on the prism. It is best to turn the prism so that the deviation has its minimum value. If the prism is slowly turned round one way and a particular line observed in the telescope the deviation of this line decreases and then stops decreasing and begins to increase. It is easy to turn the prism so that the deviation is a minimum. The angle of minimum deviation of any particular spectrum line can be found by removing the prism and turning the telescope so that it is in line with the collimator and the image of the slit coincides with the cross wire. The position of the telescope is then read off with the vernier. The prism is then put up and the telescope is turned so that the cross wire coincides with the line at the position of minimum deviation. The angle which the telescope has been turned through is the minimum angle of deviation. The refractive index of the prism for the light in question can be calculated from the angle of minimum deviation and the refracting angle of the prism. The angle of the prism can be found by turning it so that its refracting angle points towards the collimator. The parallel beam from the collimator is then divided into two parts by the edge of the prism and some of each is reflected from the two sides of the prism. The angle between the two reflected beams is measured with the telescope and, as is easily seen, it is twice the refracting angle of the prism.

The wave lengths in air of the light of the spectrum lines of different elements have been determined experimentally and tables giving the wave lengths of all known lines in the spectra of each element can be obtained. If the minimum deviations of a number of lines of known wave lengths are found, a curve can be drawn on squared paper through points the coordinates of which are the minimum deviations and corresponding wave lengths. This curve shows the relation between the minimum deviation and the wave length. If the minimum deviation of an unknown line is measured the corresponding wave length can be found from the curve and then the tables of wave lengths can be searched until a line of equal wave length is found. There may be many lines of nearly equal wave length, so that it is not always easy to be sure what element is emitting the light in question. But if the presence of any particular element is suggested, then the other lines in its spectrum can be looked for, and if many of them are found the presence of this element in the source of the light is established. The following table gives the wave lengths of some spectrum lines due to different elements. The lines selected are the brightest and most characteristic lines.

Element	Wave length in millionths of a millimetre	Colour
Barium	553·569	Yellow
Caesium	455·544	Blue
"	459·334	Blue
Calcium	558·896	Yellow
Helium	587·581	Yellow
"	587·615	Yellow
Cadmium	479·991	Blue
"	508·582	Green
"	643·847	Orange
Potassium	766·854	Red
"	770·192	Red
Lithium	670·82	Red

Element	Wave length in millionths of a millimetre	Colour
Magnesium	516·755	Green
„	517·287	Green
„	518·384	Green
Sodium	589·019	Yellow
„	589·616	Yellow
Mercury	546·097	Green
Rubidium	629·87	Orange
Strontium	460·752	Blue
„	548·115	Yellow
„	640·865	Red
Thallium	535·065	Green
Hydrogen	410·185	Violet
„	434·066	Blue
„	486·149	Blue
„	656·304	Red

Some of the lines in the spectra of the alkali metals and the alkaline earth metals can be obtained by volatilizing salts of these metals in a bunsen flame. The salt may be fused into a small bead on a loop of platinum wire and held in the flame. If electric sparks are passed between poles made of any metal, then the spectrum of the light given out by the spark contains many of the lines due to the metal and also those of the gas surrounding the poles. The yellow light emitted by sodium vapour is often used in experiments on light when light of a known wave length is required. Strong sodium light can be obtained by wrapping a piece of asbestos paper, soaked in strong sodium carbonate solution, round the top of a bunsen burner so that the flame burns at the end of the tube of paper. A mercury vapour electric lamp is an excellent source to use when strong light of known wave length is required. Light which has a definite wave length like that in a spectrum line is called monochromatic light.

A very convenient form of spectrometer for examining spectra and measuring the wave lengths of the lines with considerable accuracy is known as Hilger's wave length spectrometer. In this spectrometer the collimator and telescope are fixed at right angles to each other and the prism is of a special shape shown in Fig. 35. The angle at A is 75° and that at D 90°. If we draw a line DEB

making an angle of 60° with DA and drop a perpendicular AE on this line, then the top of the prism is divided by DB and AE into two similar triangles AED and DBC having angles of 30°, 60° and 90°, and a triangle AEB having angles 45°, 45° and 90°. A ray of light like RPS, which enters the prism through AD and is refracted so that it becomes parallel to DB, meets AB at an angle of 45° and is totally reflected through a right angle so that it is then parallel to BC and is incident on DC at an angle equal to 30° and passes out through DC in a direction at right angles to

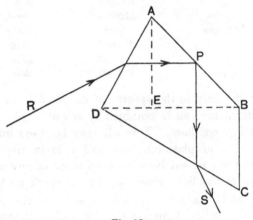

Fig. 35.

its original direction before entering the prism. Since the telescope and collimator are at right angles to each other, the rays forming an image of the slit on the cross wire in the telescope must have passed through the prism like the ray RPS. The angle of refraction at AD is equal to 30°, so if i is the angle of incidence of such a ray on AD, then $\mu = \dfrac{\sin i}{\sin 30°} = 2 \sin i$, where μ is the refractive index of the prism for light of the particular wave length in question. By turning the prism round, the angle of incidence of the light from the collimator on it can be varied, and so light of any wave length can be made to form an image on the cross wire. The prism is mounted so that it can be rotated by turning a micrometer screw. This screw carries a cylinder on which a spiral scale of wave lengths is marked, so that the wave

length of a line can be read off directly if it is brought on to the
cross wire in the telescope by turning the screw. The spiral scale
is graduated by observing a series of lines of known wave lengths.
If the lens at the end of the telescope where the spectrum is
observed, or the eye piece as it is called, is removed and a screen
with a slit in it put in the telescope so that the slit is in the
position usually occupied by the cross wire, then the slit allows
only light of the wave length indicated on the drum to pass
through it. If the collimator slit is then illuminated with white
light the other slit can be used as a source of light of any desired
wave length. An instrument which enables light of any desired
wave length to be selected from a spectrum in the way just
described is called a monochromatic illuminator.

If white light from an incandescent electric lamp or other
source is focused on the slit of a spectrometer
we get a continuous spectrum which can be
observed in the telescope. If a layer of any substance is put
up in front of the slit so that the white light has to pass through
it, then the substance may absorb parts of the white light so that
parts of the spectrum may be diminished in intensity or completely
absent. For example, a piece of red glass stops all the light except
some of the red, so that the spectrum of white light after passing
through red glass consists of a red band. Such a spectrum is
called an absorption spectrum. The absorption spectrum of a
vapour, which when hot emits light the spectrum of which is
a line spectrum, is found to consist of the continuous spectrum
crossed by sharp dark lines. The vapour absorbs chiefly light of
definite wave lengths and allows that of all the other wave lengths
to pass freely. The wave lengths of the light absorbed are found to
be the same as the wave lengths of the light which the vapour emits
when hot enough. For example sodium vapour in a bunsen flame
emits yellow light of wave lengths 589·0 and 589·6 millionths of
a millimetre. The spectrum of this light consists of two yellow
lines very close together. If white light is passed through sodium
vapour it is found that its spectrum contains two dark lines
corresponding to the wave lengths 589·0 and 589·6. This can
be shown by projecting the spectrum of an electric arc on to
a screen, with the apparatus described at the beginning of this

Absorption Spectra.

chapter, and placing a large bunsen flame filled with sodium vapour in front of the prism so that the light has to pass through the flame. The spectrum is then seen to be crossed by a dark line in the yellow. The two dark lines are so near together that they overlap and appear like one unless a very narrow slit is used. If some sodium salt is then dropped into the electric arc it gives out intense yellow sodium light and the dark line across the spectrum produced by absorption in the flame becomes highly illuminated. This shows that the sodium flame absorbs the same light which sodium vapour emits when hot. It is found that in many cases the power of a substance to emit light of any wave length is proportional to its power of absorbing the same kind of light. This is called Kirchhoff's law.

The spectrum of sunlight can be examined with a spectrometer by reflecting a beam of it on to the collimator slit through a convex lens so that the lens forms an image of the sun on the slit. It is found that the spectrum is not continuous like that of the light from an electric arc. It may be described as a continuous spectrum crossed by dark lines. The dark lines show that certain wave lengths are absent or of comparatively small intensity. These lines were first carefully studied by Fraunhofer in the years 1814–15 and they are usually called Fraunhofer's lines. It was found by Bunsen and Kirchhoff that many of the wave lengths of the light corresponding to these dark lines are the same as the wave lengths of the light in the bright lines in the spectra of the chemical elements known on the earth. For example, there is a pair of dark lines in the yellow of the solar spectrum which exactly coincide with the two yellow lines in the spectrum of the light emitted by a flame containing sodium vapour. The dark lines in the solar spectrum are due to the absorption of definite wave lengths from the white light emitted by the more central portions of the sun. The central portions are surrounded by clouds of vapour which absorb light so that the solar spectrum is the absorption spectrum of these clouds of vapour. By comparing the dark lines in the solar spectrum with the bright lines in the spectra of known elements we can tell what elements are present in the clouds of vapour round the central portions of the sun. Many of the known chemical elements have been found to be present in the sun in this way. The light from many stars gives a spectrum similar to the solar spectrum, so that in the same way it is possible to establish the presence of known chemical elements in the stars. Nebulae give bright line spectra which show that they consist of masses of glowing gases.

The most conspicuous Fraunhofer lines in the solar spectrum are usually designated by the letters H, G, F, E, D, C, B, A and correspond to light of the following wave lengths.

Line	Colour	Wave length in millionths of a millimetre	Element
H	Violet	396·8625	Calcium
G	Blue	430·8081	Iron
F	Blue	486·1527	Hydrogen
E	Green	526·9723	Iron
D	Yellow	589·6155	Sodium
C	Orange	656·3045	Hydrogen
B	Red	686·7457	Oxygen
A	Red	766·1	—

The line B is due to absorption of light by the oxygen in the earth's atmosphere.

The following table gives some values of the refractive indices of three different kinds of glass with respect to air. These glasses are named after their makers.

Wave Length of Light	Light Crown Glass (Steinheil III)	Flint Glass (Merz V)	Flint Glass (Hoffmann I)
H 396·7	1·531	1·669	1·751
G 430·8	1·527	1·658	1·736
F 486·2	1·521	1·646	1·721
D 589·6	1·515	1·633	1·704
C 656·3	1·513	1·628	1·697
A 766·1	1·510	1·622	1·690
Percentage of PbO	8·4	41·5	53·9
Percentage of K_2O	5·4	9·6	6·1

Glass consists of a mixture of silicates of potassium, sodium, lead, calcium and sometimes other elements. Crown glass contains only a small amount of lead oxide, usually about 8 % with 5 to 20 % of potassium oxide, while flint glasses contain from 30 to 60 % of lead oxide. The lead oxide is found to increase the refractive index. The length of the spectrum produced by

a prism depends on the difference between its refractive indices
for red and violet light. If μ_H denotes the refractive index of
a specimen of glass for violet light of wave length 396·7 millionths
of a millimetre, μ_A that for light of wave length 766·1 millionths
of a millimetre and μ_D that for yellow light of wave length 589·6
millionths of a millimetre, then

$$\frac{\mu_H - \mu_A}{\mu_D - 1}$$

is usually called the dispersive power of the glass for the Fraunhofer
lines H and A. The dispersive power for the lines F and C is
equal to

$$\frac{\mu_F - \mu_C}{\mu_D - 1}.$$

The extreme red and the violet parts of the spectrum of white
light are of feeble intensity, so that the dispersive power for the
blue line F and orange line C is of much greater practical importance
than that for the lines H and A.

The dispersive powers of the three glasses the refractive indices
of which are given above for the lines F and C are as follows:

	Crown (Steinheil III)	Flint (Merz V)	Flint (Hoffmann I)
$\mu_F - \mu_C$	0·008	0·018	0·024
$\mu_D - 1$	0·515	0·633	0·704
Dispersive power	0·015	0·029	0·034

The dispersive power of flint glass is about double that of
crown glass. By combining together crown glass and flint glass
prisms it is possible to make compound prisms which produce
dispersion but little deviation and are called direct vision prisms.
Compound prisms which produce little or no dispersion but
considerable deviation can also be made and are called achromatic
prisms because they do not separate the colours in white light.

The deviation produced by a small-angled prism is equal to
$(\mu - 1)\,\phi$ where μ is its refractive index and ϕ its refracting angle.
Suppose we wish to design a small-angled direct vision prism
made out of the glasses Steinheil III and Merz V that shall not
deviate D light. Let ϕ denote the refracting angle of the crown
glass and ϕ' that of the flint glass component. Then we must
have

$$0·515\,\phi + 0·633\,\phi' = 0.$$

If $\phi = 10°$ this gives $\phi' = -8.12°$.

Such a prism is shown in Fig. 36. This prism does not deviate D light but it deviates H light through the angle

$$0.669 \times 8.12 - 0.531 \times 10 = 0.12°$$

and A light through the angle

$$0.622 \times 8.12 - 0.510 \times 10 = -0.05°.$$

Thus the violet and red rays are separated by an angle $0.17°$. By combining together three crown glass prisms with two of flint

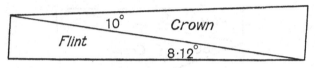

Fig. 36.

glass, direct vision prisms that do not deviate yellow light but separate violet and red rays considerably can be obtained. Such a compound prism is shown in Fig. 37. These prisms are used in

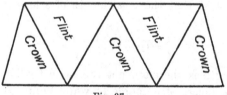

Fig. 37.

direct vision spectroscopes in which the collimator and telescope are in the same straight line, which is convenient.

To design a small-angled achromatic prism we have to arrange to neutralize the dispersion. We may make the deviation of the F light equal to that of the C light. If ϕ is the angle of the crown glass prism and ϕ' that of the flint glass one, then for the glasses Steinheil III and Hoffmann I this requires

$$0.521\phi + 0.721\phi' = 0.513\phi + 0.697\phi',$$

which gives $\phi = -3\phi'$. If $\phi' = 10°$ then ϕ must be about $30°$. Such a prism is shown in Fig. 38. This prism deviates either F or C light through an angle given approximately by the equations

$$\theta_F = 0.521 \times 30 - 0.721 \times 10 = 8.42°,$$
$$\theta_C = 0.513 \times 30 - 0.697 \times 10 = 8.42°.$$

The deviations of H, D and A lights produced by this prism are approximately as follows :

$$\theta_H = 0{\cdot}531 \times 30 - 0{\cdot}751 \times 10 = 8{\cdot}42°,$$
$$\theta_D = 0{\cdot}515 \times 30 - 0{\cdot}704 \times 10 = 8{\cdot}41°,$$
$$\theta_A = 0{\cdot}510 \times 30 - 0{\cdot}690 \times 10 = 8{\cdot}40°.$$

It appears that the deviation of the violet H light is slightly greater than the red A light, so that the prism is not perfectly achromatic for the whole spectrum.

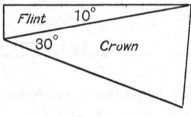

Fig. 38.

We have seen that the focal length of a lens is given by the equation

$$\frac{1}{f} = (\mu - 1) \left(\frac{1}{r} - \frac{1}{r'} \right).$$

The focal length is therefore not the same for light of different colours. The image of a source of white light formed by a lens consists therefore of a series of coloured images. The violet image is nearest the lens and the red image furthest from it. For example, suppose a convex lens of the flint glass Hoffmann I has radii of curvature $r = 200$ and $r' = -200$ cms. Its focal lengths are given by the equation

$$f = \frac{100}{\mu - 1}$$

which gives the following values :

Light	Focal Length
H	133·2 cms.
G	135·9 ,,
F	138·7 ,,
D	142·0 ,,
C	143·5 ,,
A	144·9 ,,

If sunlight is allowed to fall normally on this lens a violet image of the sun is formed 133·2 cms. from the lens and a red image 144·9 cms. from it with a series of other images in between. If a screen with a very small hole in it is put up so that one of the coloured images falls on the hole, the hole will let the light forming this image through but will stop nearly all the rest of the light.

The light getting through will therefore be red when the screen is 144·9 cms. from the lens and violet when it is 133·2 cms. from it.

The lenses used in optical instruments like telescopes and microscopes would be almost useless if they formed such a series of coloured images as that just described.

It is necessary to use compound lenses made by combining together two or more lenses of different kinds of glass so that they form images in the same position with light of different colours. Such lenses are called achromatic lenses. A lens deviates a ray of light falling on it like a small-angled prism, so that flint and crown glass lenses can be combined to produce achromatic lenses in the same way as flint and crown glass prisms can be combined into achromatic prisms. If f_C is the focal length of a crown glass lens for C light, f_F that for F light, f_C' that of a flint glass lens for C light and f_F' that for F light, then if the two lenses together form an achromatic lens we must have

$$\frac{1}{f_C} + \frac{1}{f_C'} = \frac{1}{f_F} + \frac{1}{f_F'},$$

for we have seen that two lenses of focal lengths f and f' when combined together act like a lens of focal length F such that

$$\frac{1}{F} = \frac{1}{f} + \frac{1}{f'}.$$

If the radii of curvature of the crown glass lens are denoted by r and r', those of the flint glass lens by r'' and r''', we get

$$(\mu_C - \mu_F)\left(\frac{1}{r} - \frac{1}{r'}\right) = (\mu_F' - \mu_C')\left(\frac{1}{r''} - \frac{1}{r'''}\right),$$

where μ_C and μ_F are the refractive indices of the crown glass lens and μ_C' and μ_F' those of the flint glass lens.

If the glasses are Steinheil III and Hoffmann I we have $\mu_F - \mu_C = 0\cdot008$, $\mu_F{}' - \mu_C{}' = 0\cdot024$, so that

$$\frac{1}{r'} - \frac{1}{r} = 3\left(\frac{1}{r''} - \frac{1}{r'''}\right).$$

Suppose that $r = 5$ cms., $r' = -5$ cms., and $r'' = -5$ cms.; the equation then gives $r''' = -15$ cms. This lens is shown in Fig. 39. The second surface of the crown glass lens and the first surface of the flint glass lens are often made with equal radii, so that they fit each other and can be cemented together if desired.

The collimator and telescope lenses in spectrometers always consist of achromatic combinations, so that lights of different colour are made parallel by the collimator and focused by the telescope in the same position.

Fig. 39.

CHAPTER VI

COLOUR

WHEN a source of white light is looked at through a plate of coloured glass it appears to be coloured. The coloured glass absorbs some of the colours in the white light more completely than the others. For example, red glass absorbs the yellow, green, blue and violet rays almost completely and is only transparent to the red rays. The colour of transparent bodies when seen by transmitted light is due to selective absorption of certain rays from the incident white light. A substance which absorbs all the rays equally so that their relative intensities are not changed by passing through it, is colourless like water.

When a beam of light falls on a mirror it is nearly all reflected in a definite direction, but ordinary rough surfaces scatter the light which falls on them in all directions. The scattering or irregular reflexion enables bodies to be seen when they are illuminated. All the light which falls on bodies is not scattered, much of it is absorbed and converted into heat in the body.

The incident light is partly reflected at the surfaces of the particles composing the rough surface and emerges from the surface after undergoing a series of reflexions and refractions. Some of it is reflected only once and some may be reflected and refracted several times before emerging.

If the particles composing the substance absorb rays of all colours equally, the scattered light will be of the same colour as the incident light, but if the particles absorb some of the colours more strongly than others the scattered light will be coloured differently to the incident light.

For example, the particles composing red paint absorb all the other colours more than red so that when red paint is illuminated by white light the scattered light consists of a mixture of white light and red light, so that the paint looks red. When two differently coloured paints are mixed the mixture has the absorbing powers of both paints. For example, yellow paint absorbs most of the red, blue and violet and scatters chiefly yellow and some green. Blue paint absorbs the red, yellow and violet and scatters blue and some green. A mixture of yellow and blue paints therefore absorbs nearly every colour except green, so that it acts as a green paint.

Some substances have the power of reflecting light of certain colours more strongly than others, so that when white light is incident on a polished surface of such a substance the reflected light is coloured. For example, gold and copper mirrors reflect red and yellow light better than blue and violet. Certain aniline dyes have this power of selective reflexion or surface colour in a marked degree. For example, a polished surface of solid cyanine scarcely reflects green light at all, but reflects all other colours fairly well. The reflected light from it is purple and is a mixture of red, some yellow, blue and violet rays. Substances which reflect a particular colour strongly in this way may absorb it strongly also. A substance which selectively reflects green may appear purple by transmitted light because it absorbs and reflects the green so that only the other colours get through it.

It is found that the colour of a mixture of lights of different wave lengths is not alone sufficient to determine even roughly the wave lengths present. For example, a mixture of blue light and yellow light in suitable proportions appears perfectly white although it contains no red, green or violet. The apparatus shown in Fig. 40 may be used to study the colours got by mixing lights of different wave lengths.

White light is focused on a slit at S and made parallel by a lens L. It then passes through a prism P and lens M which form a spectrum at a screen K. This screen contains an aperture equal to the spectrum; it serves to cut off any stray light. A lens N is adjusted so that it forms an image of the lens M on a white screen P. This image is a circular patch of white light in which the different colours separated at K are recombined. By covering parts of the aperture at K with suitable strips of metal any desired parts of the spectrum can be stopped and the rest let through. If we stop all the spectrum except one colour, then the circular patch on the screen at P appears of that colour. If we let through two colours at K we can observe the effect of mixing them at P. With red and green it is found that the mixture appears yellow or greenish yellow according to the proportions of red and green let through. With green and violet the mixture appears greenish blue, blue or violet. With red and violet or blue we get purple. With yellow and blue we get yellowish white, white or bluish white.

Red, green and violet cannot be obtained by mixing other colours, while yellow and blue can be so obtained. Red, green and violet are therefore called the primary colours.

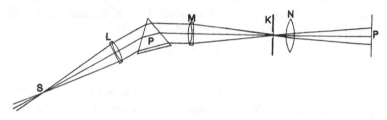

Fig. 40.

A mixture of yellow and blue appears white to the eye, but of course if it is examined with a spectroscope it is found to consist merely of yellow and blue rays, so that it is entirely different from the white light emitted by a white-hot body. It is found that any colour can be produced by mixing red, green and violet lights in proper proportions. For example, brown can be got by mixing red with a little green and violet.

Two coloured lights which when mixed give white light are said to be of complementary colours. With the apparatus just described if we cut out any colour or colours we get the complementary colour on the screen. Red and green-blue, yellow and blue, green and purple are the chief pairs of complementary colours.

REFERENCES

Physical Optics, R. W. Wood.
Colour Measurement and Mixture, Abney.

CHAPTER VII

OPTICAL INSTRUMENTS

A PHOTOGRAPHIC camera consists essentially of a convex lens
The Photographic through which light enters a box on the opposite
Camera. side of which is a screen. The distance of the
lens from the screen is adjusted so that the image of external
objects formed by the lens falls on the screen. The screen is a plane
and it is necessary that the lens shall form a distinct image all
over the surface of the screen. The formula

$$\frac{1}{f} = \frac{1}{u} - \frac{1}{v}$$

which gives the distance of the image from a thin convex lens
is only applicable when the object, its image and all the rays
of light passing through the lens are very near to the axis of the
lens.

Fig. 41 shows the shape of the image of a straight line
$ABCDE$ perpendicular to the axis of a simple convex lens LM.

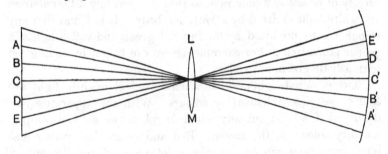

Fig. 41.

The image of C is at C', that of A at A' and so on. The image is
a curved line. The lens used in a camera has to be designed so

that it forms an image which is straight and perpendicular to the axis CC' so that it lies on the screen in the camera. The camera lens also must be an achromatic combination. Camera lenses usually consist of two or more achromatic lenses, designed so that they produce a plane undistorted image. In order to let a large amount of light into the camera in a short time the diameter of the lens has to be large. This requires rays of light to be used which do not pass through the lens near to its axis. A simple convex lens does not cause rays which pass through it at any great distance from its axis to come to a focus at the same point as rays close to the axis ; so for this reason also a more complex system than a simple thin lens is required in a camera. The theory of the design of lenses for special purposes is too complicated to be discussed in this book.

The human eye is in principle not unlike a photographic camera.
The Eye. It consists of a nearly spherical ball filled with a transparent medium composed chiefly of water. The skin of the ball is opaque, except a circular patch in front called the pupil of the eye. Light enters the eye through the pupil and is refracted at the outer spherical surface and within the eye so that rays coming from a distant point are brought to a focus on the inner surface of the skin at the back of the eye. This inner surface is called the retina, and the light forms an inverted real image of outside objects on it, just as an inverted image is formed on the screen at the back of a photographic camera. The retina is covered with an immense number of nerve endings from which nerves lead to the brain. The sensation of sight is due to the action of the light on these nerve endings in the retina which causes nervous impulses to be transmitted from the nerve endings through the nerves to the brain.

Near the front of the eye-ball, embedded in the transparent medium, is a convex lens-shaped transparent body, which has a slightly greater refractive index than the rest of the eye. The light has to pass through this lens and is made to converge slightly more in doing so. The curvature of the surfaces of this lens can be varied by the contraction of an annular muscle which surrounds it. In this way the eye can change its focal length so that sharp images of external objects at different distances can be formed on the retina. Just in front of the lens there is an opaque diaphragm with a circular aperture in the middle called the iris. This diaphragm expands and contracts so that the diameter of the aperture varies. In this way the amount of light entering the eye is regulated. In bright light the aperture becomes very small and in faint light it gets much larger.

Owing to optical defects eyes are often unable to produce distinct or sharply focused images of external objects on the retina. Such defects can often be remedied by using spectacles, which consist simply of a pair of thin lenses which are supported one in front of each eye. The curvatures of the surfaces of the lenses should be designed so that they counteract the defects of the eye and enable it to form a sharply focused image on the retina. For example it often happens that people can see objects very near to their eyes distinctly, but objects at a distance appear blurred and indistinct. This shows that the focal length of the eye, or rather its range of possible focal lengths, is too short, so that the image of a distant object is formed in front of the retina instead of on it. The focal length can be increased by using concave lenses as spectacles. Another common case is that of people who can see distant objects distinctly but not near objects. This shows that the focal lengths of their eyes are too long so that the image of near objects lies behind the retina. This defect is remedied by means of convex lenses. There are many other possible optical defects of the eye which can be remedied by properly designed spectacles. It often happens that the lens required for one eye is different from that required for the other. Spectacles should always be prescribed by a specialist, because much harm to the eyes may be done by using unsuitable lenses.

People can usually see objects most distinctly at a distance of about ten inches from the eye. This is called the normal distance of most distinct vision. Fig. 42 shows how the rays of light from an object form an image on the retina of the eye. The refraction takes place almost entirely at the front surface CD of the eye. $A'B'$ is the image of AB. The bundle of rays from any point like A on the object enters the pupil as a nearly parallel beam. The diameter of the aperture in the iris is small compared with the distance of the object from the eye, so that the angle CAD is very small. The eye may be said to be an instrument for examining parallel or very slightly diverging bundles of rays. Bundles of rays which diverge strongly or which converge cannot be brought to a focus on the retina by the eye alone.

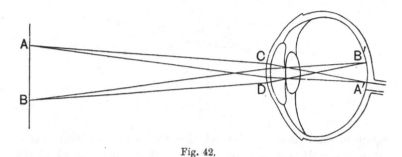

Fig. 42.

The size of the image on the retina depends on the distance
The Magnifying Glass.   of the object from the eye. In Fig. 42 it is easy to see that $A'B'$ is proportional to AB and inversely proportional to AC.

Thus the length of the image on the retina is inversely proportional to the distance of the object from the eye. If a small object is brought very near to the eye it seems larger, but it cannot be seen distinctly because its image lies behind the retina.

If a convex lens is put in front of the eye and close to the pupil, then the focal length of the combination of the lens and eye is less than that of the eye alone, so that a distinct image of an object can be formed on the retina with the object much nearer to the eye than without the lens. In this way a larger image of the object is obtained on the retina, so that the object appears to be magnified. A convex lens used in this way is often called a magnifying glass. Fig. 43 shows how the image on the retina is formed in this case.

The convex lens or magnifying glass converts the diverging bundles of rays from each point on the object into parallel or very slightly diverging bundles which the eye can focus on its retina. The lens forms a virtual image of the object which is far enough

away from the eye to be seen distinctly. The object and lens
should be placed so that the image seen appears as distinct as
possible. The distance of the virtual image from the eye is then

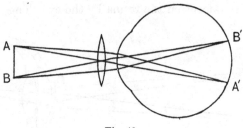

Fig. 43.

equal to the distance of most distinct vision, which is about
10 inches or 25 cms. for a normal eye. If the focal length of the
magnifying glass is f we have

$$\frac{1}{f} = \frac{1}{u} - \frac{1}{25},$$

where u is the distance of the object from the lens. The lens is
supposed to be close to the eye. If $f = 3$ cms. we get

$$\frac{1}{3} = \frac{1}{u} - \frac{1}{25},$$

which gives $u = 2\cdot7$ cms. The formation of the virtual image seen
by the eye in this case is shown in Fig. 44. Three rays are shown

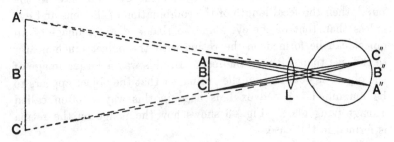

Fig. 44.

coming from the top of the object A. These are refracted by the
lens so that they meet at the top of the virtual image A' when

produced backwards. They are refracted by the eye so that they meet on the retina at A''. LB' is equal to about 25 cms. The magnifying power of the lens is the ratio of the height of the virtual image $A'B'C'$ to the height of the object ABC.

This ratio is equal to LB'/LB, or since LB is only slightly less than the focal length of the lens the magnifying power is approximately equal to $25/f$. For example, a lens of focal length 2·5 cms. magnifies about 10 times. A good magnifying glass consists of a properly designed achromatic combination of two or more lenses which act like a single lens.

The Microscope. A magnifying glass with a focal length of less than about 1 cm. is inconvenient so that when it is desired to obtain a higher magnifying power than about 25 a simple magnifying glass is not suitable. The microscope is the instrument used when high magnifying powers are required. It consists essentially of two lenses or systems of lenses called the objective and the eye-piece.

The objective is a convex lens of short focal length, which forms a real inverted image of the object. This real image is viewed through the eye-piece, which acts as a simple magnifying glass and forms a virtual image of the real image at the distance from the eye of most distinct vision. The objective is placed at a distance from the object only slightly greater than its focal length so that the real image is larger than the object.

Let f be the focal length of the objective, u the distance of the object from it and v the distance of the real image from the objective. We have

$$\frac{1}{f} = \frac{1}{u} - \frac{1}{v}.$$

For example, suppose $f = 1$ cm. and $v = -15$ cms. This gives $\frac{1}{1} = \frac{1}{u} + \frac{1}{15}$ or $u = \frac{15}{14}$ cms. In this case the object is $\frac{15}{14}$ cms. from the objective and the real image is 15 cms. from it on the opposite side. The height of the real image is to the height of the object in the ratio $-\frac{v}{u}$ or in the particular case considered $15 \div \frac{15}{14}$, so that the real image in this case is 14 times as high as the object. Let f' denote the focal length of the eye-piece. Its magnifying power is approximately equal to $25/f'$ so that the total magnifying power of the microscope is equal to

$$-\frac{v}{u} \times \frac{25}{f'}.$$

If $v = -15$ cms., which is about the value usually adopted, and if f is small, then u and f are nearly equal so that the magnifying power is nearly equal to

$$\frac{15 \times 25}{ff'} = \frac{375}{ff'}.$$

For example, if $f = 0\cdot5$ cm. and $f' = 3$ cms. the magnifying power is about 250. Fig. 45 shows how the real and virtual images are

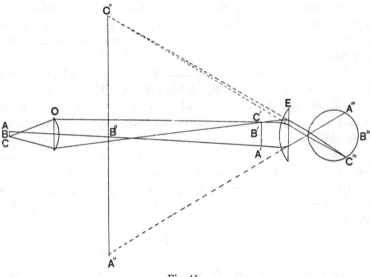

Fig. 45.

formed in a microscope. ABC is the small object. Two rays from C are shown that are refracted by the objective O and form a real image of C at C'. These rays then fall on the eye-piece E and are refracted so that when produced backwards they meet at the virtual image C'', which is 25 cms. from the eye. The two rays are refracted at the surface of the eye and form a real image of C on the retina at C''' as shown.

The objectives used in microscopes are not simple thin lenses but are combinations of several lenses designed to be achromatic and to produce brightly illuminated undistorted images. The design of a good objective of short focus is a very complicated problem.

Telescopes. Telescopes are used for viewing distant objects, and produce a magnified image. They are very similar in their mode of action to microscopes. A real image of the distant object is formed by a convex lens or a concave mirror and this image is viewed through an eye-piece which magnifies it. If a mirror is used to form the real image the telescope is called a reflector, and if a lens is used it is called the object-glass and the telescope a refractor or a refracting telescope. The magnifying power of a telescope is taken to be the angle subtended at the eye by the image seen divided by the angle subtended by the object.

Fig. 46 shows the path of three rays through a reflecting telescope into the eye. The three rays, which are practically parallel, are

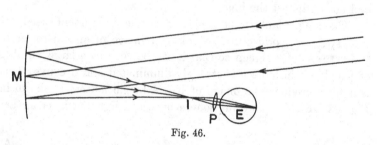

Fig. 46.

reflected from the concave mirror M so that they form a real image of the distant point at I. They then pass through the eye-piece P and are brought to a focus on the retina of the eye E. The distance MI is equal to the focal length f of the mirror M. The height of the real image at I is therefore equal to $f\theta$, where θ is the angle which the distant object subtends at M. The eye-piece forms a virtual image of the real image and its magnifying power is equal to $25/f'$ where f' is its focal length. The virtual image therefore subtends an angle at the eye equal to $f\theta \times \dfrac{25}{f'} \div 25$, and the magnifying power of the telescope is equal to this divided by θ, so that it is equal to $f \div f'$. To get a large magnifying power it is therefore necessary to use a mirror of long focal length and an eye-piece of short focal length.

A refractor is shown in Fig. 47. It works in the same way as a reflector and its magnifying power is equal to f/f', where f is the focal length of the object-glass and f' that of the eye-piece.

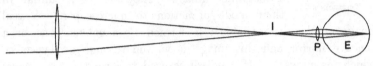

Fig. 47.

The object-glass of a telescope is always an achromatic lens, and the curvatures of its surface should be designed so that it may produce an undistorted distinct image of a distant object. The design of a good object-glass is a complicated problem, which will not be discussed in this book. The eye-piece usually consists of a combination of two or more lenses designed so as to be achromatic and not to distort the image.

The magic lantern or projector is the instrument used for **The Magic Lantern.** projecting an enlarged image of an object on to a screen so that it can be seen by a large number of people. The object is strongly illuminated and a convex lens is used to produce the image of it. A magic lantern is shown in Fig. 48. S is a powerful source of white light such as an arc

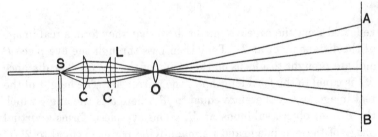

Fig. 48.

lamp. C and C' are two large convex lenses which focus the rays from S on to a convex lens O. The rays drawn pass through the lens O at its centre, so that they are not deviated by this lens.

The light passing through the lens O illuminates the white screen AB. If a partly transparent object like a lantern slide is put up at L in front of the lens C' the position of the lens O can be adjusted so that it produces a sharply focused image of the object on the screen. Fig. 49 shows how this image is formed.

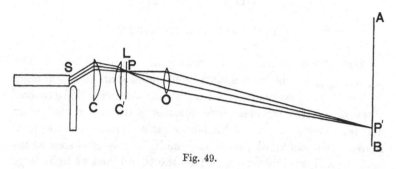

Fig. 49.

Three rays are shown starting from different points on the source S which meet at P on the object and are focused by the lens O on the screen at P'. Thus P' is a real image of P. The middle ray of the three passes through O at its centre, so it is like one of the rays drawn in the previous figure. The lenses C and C' form a real image of the source S at or near to the lens O. The lens O is usually replaced by an achromatic combination of two or more lenses designed so as to produce an undistorted image sharply focused all over the screen.

REFERENCE

The Theory of Optical Instruments, E. T. Whittaker.

CHAPTER VIII

THE VELOCITY OF LIGHT

GALILEO attempted to determine the velocity of light. Two observers situated some miles apart on a clear night were provided with lanterns which could be covered when desired by screens. One observer A uncovered his lantern and the other B uncovered his lantern as soon as he saw the light from A. It was found that A saw the light from B as soon as the lantern at A was uncovered, so that the time taken by light to go from A to B and back was too small to be measured in this way. If the distance from A to B and back was 10 miles and the smallest time interval which could have been detected was, say, 0·1 second, such an experiment would show that the velocity of light was not less than 100 miles per second.

The Velocity of Light.

In 1676 Roemer, a Danish astronomer working in the Paris Observatory, made the first estimate of the velocity of light. The planet Jupiter has several satellites or moons which revolve round it in nearly circular orbits with nearly uniform velocity. The satellite nearest to Jupiter goes behind Jupiter as seen from the earth once in each revolution. Roemer observed the times at which this satellite disappeared behind Jupiter during more than a whole year. The average interval between successive disappearances is 42 hours 28 minutes 36 secs. Thus we may regard the satellite and Jupiter as sending out a series of signals at equal intervals of time. Roemer found that the observed time intervals between successive disappearances were not equal but varied throughout the year in a regular way. While the earth was moving away from Jupiter the intervals were longer than the average and while it was moving towards Jupiter they were shorter than the average. Roemer attributed this to the finite velocity of light and calculated this velocity from his results.

The earth moves round the sun in a nearly circular orbit of radius about 1.5×10^{13} cms. During a year Jupiter only moves round the sun through a small part of his orbit which lies nearly in the same plane as that of the earth. Let us suppose that the times at which the satellite disappears are observed when the sun, earth and Jupiter are in a straight line, with the earth between the sun and Jupiter, and that the observations are continued until the earth is again in this position. The distance between the earth and Jupiter will then be the same at the start as at the end of the series of observations, but at the middle of the series this distance will be greater by the diameter of the earth's orbit or 3×10^{13} cms. Suppose that the total number of intervals observed is $2n$. Roemer found that the first half of the intervals, that is, the first n intervals as observed on the earth, occupied about 2000 secs. longer than the second n intervals. That is, the nth disappearance, which happens when the earth is at the greatest distance from Jupiter, is 1000 secs. late. Roemer explained this by supposing that the light took 1000 seconds to travel the extra distance, which gives for the velocity of light in empty space $\dfrac{3 \times 10^{13}}{1000}$ or 3×10^{10} cms. per sec.

The diameter of the earth's orbit was not accurately known in Roemer's time, so that his estimate was not very exact. The correct value has been used in the calculation just given.

Another method of getting the velocity of light by astronomical observations was discovered by Bradley in 1726. He observed the angular positions of different stars relative to the earth during long periods of time and found that they varied slightly in a periodical manner with a period of one year. Consider for example a star in a direction perpendicular to the plane of the earth's orbit round the sun. Bradley found that such a star appears to describe a small circle the angular diameter of which is 41″.

In Fig. 50, let BA represent the velocity of light from a star and BC that of the earth, both relative to the sun. The velocity of the light relative to the earth is represented by CA, so that the motion of the earth causes a change in the direction of motion of the light as seen by an observer on the earth. As the earth goes round the sun the direction of its motion changes and after six months it is reversed, so that then the direction of the light is changed in the opposite direction.

The angular diameter of the small circle described by the star is therefore equal to twice the angle BAC so that we have

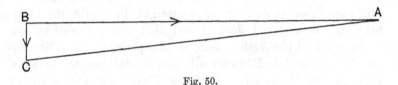

Fig. 50.

$B\hat{A}C = 20\cdot5''$. If C denotes the velocity of light and v that of the earth in its orbit, then

$$\tan B\hat{A}C = \frac{v}{C} = \tan 20\cdot5'' = \frac{1}{10000}.$$

The velocity of the earth in its orbit round the sun is equal to the circumference $2\pi \times 1\cdot5 \times 10^{13}$ cms. divided by the number of seconds in a year, which is nearly $3\cdot16 \times 10^7$. This gives $v = 3 \times 10^6$ cms. per sec., so that $C = 10^4 \times 3 \times 10^6 = 3 \times 10^{10}$ cms. per sec. Thus Bradley's method gives practically the same result as Roemer's.

The velocity of light in a vacuum, 3×10^{10} cms. per sec. or 186,413 miles per second, is so great that it is difficult to measure it on the earth. Two methods however have been used successfully. The first method was used by Fizeau in 1849. Fig. 51 is a diagram of his apparatus. S is a source of light the rays from which pass through a convex lens L and are reflected from a plane mirror M so that they form an image of S at F. They then pass through a lens L' which makes them parallel. The parallel beam passes through a lens L'' which causes it to converge to a focus on the surface of a plane mirror C, which reflects the rays back through L'' and L' so that they form a second image of the source at F. The distance between L' and L'' in Fizeau's experiment was about 866,300 cms. This long distance is omitted in the figure. The rays then fall again on the mirror M. This mirror is a glass plate the surface of which is coated with a layer of silver so thin that it only reflects about half the light which falls on it and allows the other half to pass through. About half the light returning from

C through L' therefore passes through M to an eye-piece L''', through which it passes into the observer's eye E where it forms an image of S on the retina at D. The eye E is drawn on a much larger scale than the rest of the apparatus. At F there is a wheel W which can be made to rotate at a known speed about the axis AA'. The circumference of this wheel is cut into a number of teeth or cogs with spaces between the cogs equal in width to the teeth. The wheel is placed so that the image at F lies close to the circumference, and as the wheel rotates the light from S is stopped by the cogs but passes through the spaces between the cogs.

If the wheel is turned slowly, the eye E sees an intermittent image of S, for each cog stops the light. If the wheel is turned

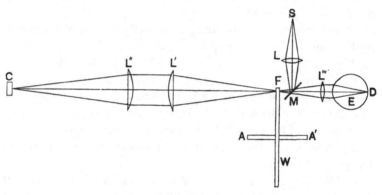

Fig. 51.

more quickly, the light which has passed through between the two cogs is stopped on its way back to E by a cog which has moved to the position of the image at F while the light travelled from F to C and back. If the speed of the wheel is then doubled the light gets through to E again, because when it gets back to F the cog has passed and the next space between two cogs is there. If the speed is made three times the first speed the light is stopped again and so on. It is possible to measure a series of speeds of the wheel proportional to the numbers 1, 2, 3, 4, 5, 6, etc. at which the light is stopped when the number is odd and gets through when it is even.

In Fizeau's experiment the wheel had 720 teeth and it was found that increasing the speed by 12·6 revolutions per second caused the image of the source seen by the observer to change from zero intensity to its maximum intensity, or increasing the speed 25·2 revolutions per second caused the brightness of the image to change from one zero of intensity to the next zero. This shows that the time taken by the light to go from F to C and back was equal to $\dfrac{1}{25\cdot2 \times 720}$ second. The distance from F to C and back was 1,732,600 cms. so that the velocity of the light worked out as

$$1732600 \times 25\cdot2 \times 720 = 3\cdot1 \times 10^{10} \text{ cms. per sec.}$$

In 1874 Cornu repeated Fizeau's experiment with improved apparatus and found the velocity of light to be almost exactly 3×10^{10} cms. per second.

Another method of measuring the velocity of light was invented by Foucault in 1850. The principle of this method is as follows. In Fig. 52, S is a source of light which passes through a lens L and is then reflected from a plane mirror M on to a concave mirror M' at the surface of which the lens L forms an image of S. The mirror M' reflects the light back so that it returns along its path to M. The mirror M is made to rotate rapidly in the direction shown by the arrow about an axis at M perpendicular to the plane of the paper.

While the light goes from M to M' and back the mirror M turns through an angle, so that the returning beam is not reflected from M back into L but in a direction like MS'. The returning light is received in a telescope $L'L''$ and forms an image of S at S', which is observed by the eye E through the eye-piece L''. The mirror M' is placed so that its centre of curvature lies on the axis of rotation of the rotating mirror M, so that as the image of S sweeps across M' it is all the time reflected back on to M. If the rotating mirror makes n revolutions per second and the angle $S'MS = \theta°$, the time taken by the light to go from M to M' and back is $\theta/(2 \times 360n)$, so that if $MM' = d$ the velocity of light is

$$2d \div \frac{\theta}{720n} = \frac{1440nd}{\theta}.$$

Foucault's own experiments were not very exact; but his experiment has been repeated with greatly improved apparatus by Michelson and later by Newcomb. Both these observers found the velocity of light in a vacuum to be very nearly equal to 3×10^{10} cms. per second.

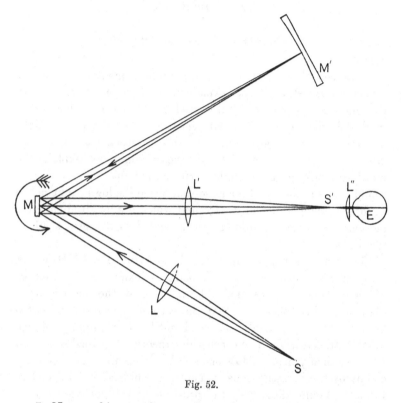

Fig. 52.

In Newcomb's experiments d was about 300,000 cms. so that θ would have been equal to about $14\cdot4°$ if n were 1000 revolutions per second, for these numbers give for the velocity

$$\frac{1440 \times 1000 \times 3 \times 10^5}{14\cdot4} \text{ or } 3 \times 10^{10} \text{ cms. per sec.}$$

By putting a tube filled with water between the mirrors M and M' Foucault showed that the velocity of light in water was about $2\cdot25 \times 10^{10}$ cms. per second, or only $\frac{3}{4}$ of its velocity in air.

24—2

CHAPTER IX

INTERFERENCE AND DIFFRACTION

THE fact that rays of light in a uniform medium like air or a vacuum are straight was explained by Newton on the theory that light consists of material particles shot out from hot bodies in all directions with very great velocities. The theory that light is a wave motion in a medium that fills all space was opposed by Newton on the ground that this theory could not explain the rectilinear propagation of light. He argued that a wave motion would travel round an obstacle, so that sharp shadows ought not to be formed, and pointed out that sound, which is a longitudinal wave motion in air, travels around corners and does not give shadows like light.

The wave theory was first put forward in 1678 by Huygens, who showed how it could explain reflexion and refraction. Huygens however was not able to explain the formation of shadows on his theory; so that the corpuscular theory supported by Newton was generally accepted until the beginning of the nineteenth century. The corpuscular theory was overthrown and the wave theory firmly established early in the nineteenth century chiefly by the investigations of Young in England and Fresnel in France. Young discovered the principle of interference between two trains of waves, and showed that certain optical phenomena could be explained by it which could not be explained satisfactorily on the corpuscular theory. Fig. 53 shows an experiment due to Young. S is a narrow slit, the length of which is perpendicular to the paper, in an opaque screen which is strongly illuminated by means of a source of white light L and a convex lens. The light from the slit S falls on a second screen containing two narrow slits A and B parallel to the slit S. The slits A and B are only a short

distance, say two or three millimetres, apart. The light from
A and B falls on a white screen CD. The distances SA and AC
between the screens may each be one or two metres. According
to the corpuscular theory we should expect to get on the screen
CD two bright lines of light in the positions got by joining SA

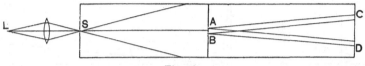

Fig. 53.

and SB by straight lines and producing these lines to meet CD.
Young found that the light from each of the slits A and B was
diffused over a considerable area of the screen CD and that where
the illumination due to A was superposed on that due to B a series
of parallel bright and dark coloured bands appeared. The bands
were parallel to the lengths of the slits. The central band at the
bisector of the angle ASB was white, and on each side of it was
a dark band and then nearly white bands. Outside these came
dark bands, and then a series of coloured bands getting rapidly
more confused and indistinct as the distance from the central white
band increased. If instead of white
light coloured light is used to illuminate the slit S, then bright bands
are obtained of the same colour as
the light used, with dark bands
between them. The bands are
further apart with red light than
with blue light. Fig. 54 shows the
appearance of the bands. They
are called interference bands.

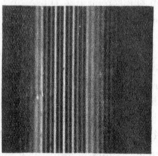

Fig. 54.
(Reproduced by permission from
Schuster's *Optics*.)

A more convenient way of
obtaining them is by means of an
instrument invented by Fresnel called Fresnel's bi-prism. This
is shown in Fig. 55. S is a slit in an opaque screen which is
illuminated from the left-hand side. The bi-prism AED has three
plane sides AD, AE and ED. The angles at A and D are equal
and very small, so that the angle at E is nearly $180°$.

The light from S after passing through the bi-prism falls on a white screen CC'. SE should be about 25 cms. and EC about 200 cms. or more. The light that passes between A and E is deviated downwards by the prism and illuminates the screen

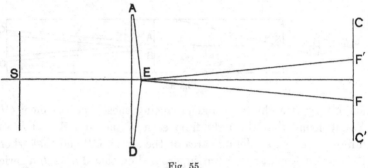

Fig. 55.

between C and F. The light that passes between E and D is deviated upwards and illuminates the screen between C' and F'. Thus between F and F' the screen is illuminated by light that has passed through AE and also by light that has passed through ED. Between F and F' a series of interference bands is formed on the screen like those obtained in Young's experiment. Much brighter bands are obtained with the bi-prism than with two slits because the prism allows much more light to pass.

The formation of such interference bands is easily explained by the wave theory of light, and the wave length can be calculated from the distance between the bands. In Fig. 56 let A and B be the two slits in Young's experiment and PQ the screen on which the interference bands are seen. Let P be a point such that $AP = BP$. At P there is a bright band. If the light from S is monochromatic light, then according to the wave theory it consists of trains of waves. These waves pass through the slits A and B and then diverge from A and B so that the screen is illuminated by two series of trains of waves. At P the two paths from S, SAP and SBP, are of equal length, so that waves reaching P from A will agree in phase with those from B. At P, therefore, the two sets of waves reinforce each other, so that there is a bright band at P. Now consider a point Q. The two series of waves which arrive at Q

travel along SAQ and SBQ respectively. The path SBQ is longer than the path SAQ by the distance $BQ - AQ$. If $BQ - AQ$ is equal to half a wave length of the light the two series of waves at Q will produce displacements in opposite directions and so will destroy each other in the same way as with trains of sound waves, so that

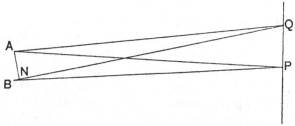

Fig. 56.

the screen will be dark at Q. Also if $BQ - AQ$ is equal to any odd number of times a half wave length, interference will take place and the screen will be dark. If $BQ - AQ$ is equal to an even number of half wave lengths, the two trains of waves reinforce each other and produce a bright band at Q.

Let D denote the perpendicular distance between the slits and the screen; let $AB = 2d$ and $PQ = x$. Then we have

$$BQ^2 = D^2 + (x + d)^2, \text{ and } AQ^2 = D^2 + (x - d)^2.$$

Hence

$$(BQ - AQ)(BQ + AQ) = 4xd \text{ so that } BQ - AQ = 4xd/(BQ + AQ).$$

If there is a bright band at Q then $BQ - AQ = n\lambda$ so that $n\lambda = 4xd/(BQ + AQ)$, where λ is the wave length of the light and n denotes any of the whole numbers 0, 1, 2, 3, 4, etc. When x is small we may put $BQ + AQ = 2D$ so that $n\lambda = 2xd/D$ very nearly. If $n = 0$ we have $x = 0$ and Q is then at the central bright band. If $n = 1$ then $x = \lambda D/2d$, and Q is at the first bright band after the central one on one side. If $n = 2$, Q is at the second bright band and so on.

The wave length of the light can be found by measuring the distance from one bright band to another one and counting the number of spaces between bright bands in between the two. Suppose for example $AP = 300$ cms., $AB = 0.3$ cm., and that yellow

light is used.   Then it will be found that ten spaces between
bright bands occupy about 0·6 cm.   Hence

$$10\lambda = \frac{0\cdot3}{300} \times 0\cdot6$$

$$\text{or } \lambda = \frac{6}{100,000} \text{ cms.}$$

This is about the wave length of yellow light.

To explain the laws of reflexion and refraction of light by the
wave theory, Huygens used a principle called after
him.   According to this principle when the ether
at any point is disturbed by the passage of a light wave over the
point, then the point becomes a centre of disturbance in the ether
and a spherical light wave diverges from it.   In Fig. 57 let S be
a source of light and ABC a spherical wave which started from S.
At every point on the surface of the sphere ABC the ether is

Huygens'
Principle.

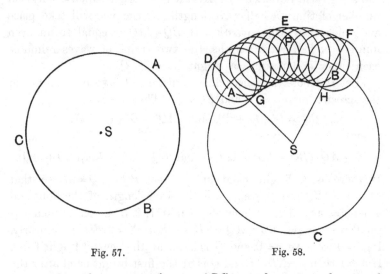

Fig. 57.   Fig. 58.

disturbed, so that every point on ABC must be supposed to send
out a spherical wave.   After a time t these waves will all have
radii equal to ct where c denotes the velocity of light.   Fig. 58
shows a number of circles of equal radii drawn with their centres
on the circle ABC between the points A and B.   All these circles
touch a circle DEF of radius $r+ct$ and a circle GH of radius

$r - ct$. Here r is the radius of the circle ABC. The wave from S is in the position AB at the time 0, and at the time t each element of it must have moved a distance ct from its position on AB and must therefore lie on the surface of the sphere of radius ct with centre at the position of the element at time 0. Huygens supposed that a surface drawn so as to touch all the spheres of radii ct would coincide with the light wave after the time t. For example, in Fig. 58, the wave between A and B after the time t would be in the position DEF. The rays of light Huygens supposed coincide with the radii of the spherical waves which end on the surface which they all touch. We might expect according to Huygens' principle that the wave AB would also produce a wave GH moving inwards, which does not happen in fact. Also it is not clear why the spherical waves produce no effect except at the surface of the sphere DEF which they all touch. Huygens' principle alone therefore does not help us very much, but Fresnel showed that by combining Young's principle of interference with Huygens' principle the nearly rectilinear propagation of light could be explained. Before discussing the elements of Fresnel's theory we may consider how Huygens explained the laws of reflexion and refraction.

Let ACB in Fig. 59 be a plane light wave and AB'' the surface

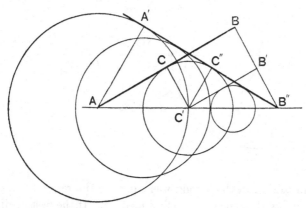

Fig. 59.

of a plane mirror. Let the wave at B move along $BB'B''$ and C along CC' so that CC' and $BB'B''$ are rays of light and are perpendicular to the wave AB. Where the wave meets the mirror at A

we suppose, following Huygens, that a spherical wave starts out
from the mirror the radius of which after a time t is ct. By the
time the wave gets to the mirror at B'' the radius of the wave
starting from A will be equal to BB''. Describe a circle with centre
A and radius equal to BB''. Also with centre C' and radius equal
to $B'B''$ describe another circle. This circle represents the spherical
wave starting from C' when the wave has got to B''. If we draw
a plane $B''C''A'$ touching the circles it will be in the position of
the reflected wave at the instant when B has got to the mirror at
B''. Since the angles at B and A' are right angles and $AA' = BB''$
it follows that AB and $B''A'$ are equally inclined to the mirror in
accordance with the laws of reflexion. $C'C''$ is the reflected ray
corresponding to CC', for it joins the centre of the spherical wave
starting from C' to the point C'' where it touches the wave surface
$A'C''B''$.

In a similar way Huygens explained the laws of refraction.
He supposed that the ratio of the velocity of light in the first

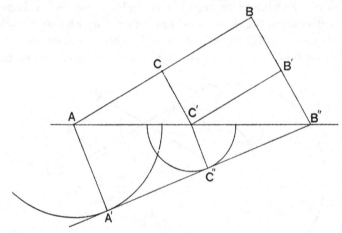

Fig. 60.

medium to that in the second was equal to the refractive index of
the second with respect to the first medium. If the first medium is
air in which the velocity of light is c and the second glass or any
other transparent substance in which the velocity of light is c', then
Huygens supposed $\mu = c/c'$, where μ is the refractive index of glass
with respect to air.

In Fig. 60 let ACB be a plane wave in air and AB'' the plane surface of a piece of glass. Let the disturbance at B move along $BB'B''$ and that at C along CC' so that CC' and BB'' are the paths of rays and are perpendicular to AB. Huygens supposed that when the wave meets the glass at A a spherical wave would start from A in the glass, the radius of which after a time t would be $c't$. By the time B gets to B'' the radius of the wave from A is therefore equal to $BB'' \times \dfrac{c'}{c}$. Describe a circle with centre A and radius equal to $BB'' \times \dfrac{c'}{c}$. In the figure AA' is taken equal to $\tfrac{3}{4}BB''$. In the same way when the wave gets to the glass at C' a wave starts from C' the radius of which when the wave has got to B'' is $B'B'' \times \dfrac{c'}{c}$. Describe a circle with centre C' and radius $B'B'' \times \dfrac{c'}{c}$. According to Huygens a plane $B''C''A'$ drawn so that it touches all the spheres represents the position of the refracted wave when the wave has got to B''. The angle of incidence i of the wave on AB'' is BAB'' and the angle of refraction r is $AB''A'$. We have

$$\sin i = \frac{BB''}{AB''} \text{ and } \sin r = \frac{AA'}{AB''},$$

so that

$$\frac{\sin i}{\sin r} = \frac{BB''}{AA'}.$$

But

$$AA' = BB'' \times \frac{c'}{c} \text{ and } \frac{c}{c'} = \mu,$$

so that we get

$$\mu = \frac{\sin i}{\sin r},$$

in agreement with the laws of refraction. According to the corpuscular theory it was supposed that the velocity of light was greater in glass or water than in air. We saw at the end of the last chapter that Foucault found the velocity in water to be only $\tfrac{3}{4}$ that in air, and the refractive index of water with respect to air is equal to $\tfrac{4}{3}$; so that Foucault's experiment agreed with Huygens' wave theory and not with the corpuscular theory.

Fresnel showed that by combining Huygens' principle with Young's theory of interference the rectilinear propagation of light could be satisfactorily explained on the wave theory. He also showed that the observed deviations from exact rectilinear propagation were in accordance with the wave theory. In this book we shall consider only one simple case in an elementary way.

In Fig. 61 let S be a small source of light and AB an opaque screen with a straight edge at A perpendicular to the plane of the paper, which throws a shadow on a plane white screen CD. The screen is illuminated by S from O to C, but from O to D the opaque

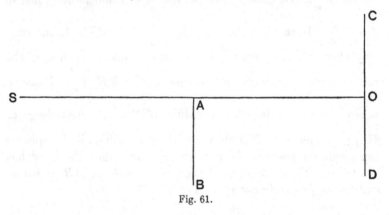

Fig. 61.

screen AB casts a shadow. If S is a narrow slit, illuminated with monochromatic light from the left side, with its length parallel to the straight edge at A and the distances SA and AO are both one or two metres, then on examining the edge of the shadow at O it is found that just above O there are narrow bright and dark bands getting rapidly narrower and less distinct as the distance above O is increased, while below O the illumination of the screen rapidly but gradually gets fainter, so that a short distance below O it is not appreciable. The appearance of the edge of the shadow is shown in Fig. 62. OO' is the level of the geometrical shadow, that is, where the straight line SAO meets the screen. Bright bands are marked BB' and dark bands DD'. According to the corpuscular theory we should have expected the screen to be uniformly bright above OO' and uniformly dark below it. The slight bending of the light into the region below OO' is called diffraction and the bands seen above OO' are called diffraction bands. The problem is to explain the bands on the wave theory and to show why the illumination extends only such a little way below OO'. Huygens could not solve this problem and the wave theory was not accepted until this was done by Fresnel.

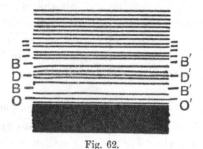

Fig. 62.

Let $ABCD$ in Fig. 63 represent a plane area and suppose that a train of plane waves of wave length λ is moving perpendicular

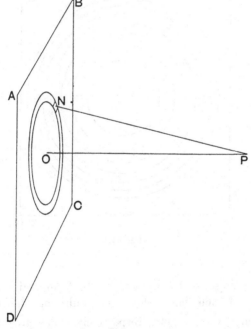

Fig. 63.

to the area so that OP which is normal to $ABCD$ is a ray of light. The wave surfaces are then planes parallel to the area $ABCD$.

The disturbance in the ether at P may be regarded as produced by the disturbances which diverge as spherical waves from every point on the area $ABCD$. The resultant effect at P at any instant can be got by finding the resultant of all the effects at P at that instant coming from all the different parts of $ABCD$. The amplitude of the vibration at P due to the train of spherical waves coming from a small area on $ABCD$ may be taken to be proportional to the small area and to diminish as its distance from O increases.

Describe a series of concentric circles on the plane $ABCD$ with centres at O and radii equal to $\sqrt{x\lambda}$, $\sqrt{2x\lambda}$, $\sqrt{3x\lambda}$ and so on. Here $x = OP$. Such a series of concentric circles is shown in Fig. 64.

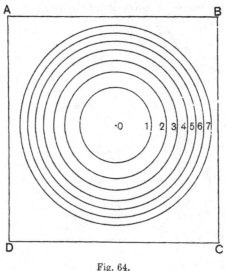

Fig. 64.

If the wave length of the light is 6×10^{-5} cm. and x is equal to $\frac{1}{6} \times 10^{5}$ or 16,666 cms., then the radii of the circles are 1 cm., $\sqrt{2}$ cms., $\sqrt{3}$ cms., etc., respectively. The area of the first circle is $\pi x\lambda$, that of the second $2\pi x\lambda$ and so on. Thus the areas between successive circles are all equal to $\pi x\lambda$. The circles therefore divide the area $ABCD$ into equal parts. These equal areas are called Fresnel's zones. The distance of the circumference of the first circle from P is equal to

$$\sqrt{x^2 + x\lambda} = x\left(1 + \frac{\lambda}{x}\right)^{\frac{1}{2}} = x + \frac{\lambda}{2},$$

since λ is very small compared with x.

The distance of the second circle from P is equal to $\sqrt{x^2 + 2x\lambda} = x + \lambda$. In the same way the distance of the nth circle from P is equal to $x + n\dfrac{\lambda}{2}$. The circles therefore divide $ABCD$ into equal areas each one of which is on the average half a wave length further from P than the one next to it and nearer to O. Let the amplitude of the disturbance at P due to the first circle be denoted by S_1, that due to the area between the first and second circles or the second area by $-S_2$, that due to the third area by S_3 and so on. The amplitudes due to even numbered areas are taken of opposite sign to those due to odd numbered areas because increasing the distance from P by half a wave length changes the sign of the disturbance at P. The total amplitude at P is therefore equal to

$$S_1 - S_2 + S_3 - S_4 + S_5 - S_6 + \ldots.$$

The areas are all equal but their distances from O increase so that S_2 is less than S_1, and S_3 less than S_2 and so on.

We suppose that S_n becomes negligible when n is very large. To find the sum of the series $S_1 - S_2 + S_3 - S_4 + \ldots$ let

$$S_1 - S_2 = \alpha_1,$$
$$S_2 - S_3 = \alpha_2,$$
$$S_3 - S_4 = \alpha_3,$$
$$\text{etc.}$$

Adding up these equations we get

$$S_1 = \alpha_1 + \alpha_2 + \alpha_3 + \ldots,$$

because when n is large S_n becomes negligible.

The series $S_1 - S_2 + S_3 - S_4 + \ldots$ is equal to $\alpha_1 + \alpha_3 + \alpha_5 + \ldots$, and since the terms in the series

$$S_1 - S_2 + S_3 - S_4 + \ldots$$

diminish slowly and regularly in magnitude we see that

$$\alpha_1 + \alpha_3 + \alpha_5 + \ldots = \alpha_2 + \alpha_4 + \alpha_6 + \ldots \text{ approximately.}$$

Hence each of these series is equal to $\frac{1}{2}S_1$, since the sum of the two
is equal to S_1.

The resultant amplitude of the vibration at P is therefore
equal to one-half that due to the disturbance coming from the first
circle described round O.

If we describe a circle with centre O and radius $\sqrt{\dfrac{x\lambda}{2}}$, that is
one of half the area of the first Fresnel zone, then we may regard
the disturbance at P as coming from this small area at O and the
disturbances from all the rest of the area $ABCD$ can be regarded
as destroying each other by interference at P and so having no
effect there.

In Fig. 65 let RP be a ray of light and AOB a surface
perpendicular to it at O. It follows from the above that the light

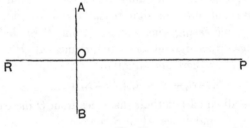

Fig. 65.

at any point on the ray, say P, may be regarded as coming from
a very small area on AOB surrounding O. Now O can be taken
anywhere along the ray RP so that it follows that the light travels
along the ray and the light at P will not be cut off by an opaque
screen unless it crosses the ray RP or comes so near to RP that it
cuts off part of the small area around the ray from which the light
at P comes. If $OP = x$ the radius of the small area at O is
$\sqrt{\dfrac{x\lambda}{2}}$. For example if $x = 100$ cms. and $\lambda = 6 \times 10^{-5}$ cm. the
radius is $\sqrt{3 \times 10^{-3}}$ or about $\frac{1}{18}$ cm.

In this way Fresnel explained the formation of shadows on the
wave theory. It is clear that the approximately rectilinear
propagation of light depends on the fact that the wave length is
so small. For example if we take $\lambda = 10$ cms. and $x = 100$ cms.

we get $\sqrt{\dfrac{x\lambda}{2}} = 22{\cdot}4$ cms., so that the illumination at P would come from an area of radius $22{\cdot}4$ cms. at O and so could not be regarded as travelling along close to the line OP.

The series $S_1 - S_2 + S_3 - S_4 + \ldots$ may be written

$$\frac{S_1}{2} + \frac{1}{2}\{(S_1 - S_2) + (S_3 - S_2) + (S_3 - S_4) + \ldots\}.$$

The terms $(S_1 - S_2)$, $(S_3 - S_2)$, etc. are all very small, for there is no reason to suppose that the effect due to a small area diminishes at all rapidly with its distance from O.

Consequently the successive rings round O may be supposed to nearly destroy each other's effects; the outside half of each ring destroys the effect due to the inside half of the next ring.

The total effect is therefore that due to the inside half of the first circle at O.

Let us now consider more in detail the effect of an opaque screen on the illumination at P. Let the opaque screen cover the whole of the area $ABCD$ (Fig. 64) except for a circular area with its centre at O. If the circular hole in the screen has a radius equal to $\sqrt{x\lambda}$ it will let through the first Fresnel zone only, so that the amplitude at P will be equal to S_1, and so will be double that when the screen is removed. The intensity of the illumination at P is proportional to the square of the amplitude of vibration, so the intensity will be four times that when the screen is removed. If the hole in the screen has a radius equal to $\sqrt{2x\lambda}$ it will let through the first two zones and the amplitude at P will be $S_1 - S_2$ which is very small. Thus increasing the radius of the hole from $\sqrt{x\lambda}$ to $\sqrt{2x\lambda}$ diminishes the illumination at P from four times the value without the screen to a very small fraction of this value. These at first sight surprising results of the theory have been verified experimentally. Such results clearly disprove the corpuscular theory of light.

If instead of a screen with a circular hole in it we use an opaque circular disk we get equally remarkable results. Suppose the centre of the disk is at O and its radius is equal to that of one of the Fresnel zones, then it cuts off some of the first terms in the series

$$S_1 - S_2 + S_3 - S_4 + \ldots$$

Suppose the first four circles are covered by the disk, then the amplitude at P is equal to

$$S_5 - S_6 + S_7 - S_8 + \dots,$$

which is equal to $\dfrac{S_5}{2}$ and so is nearly equal to $\dfrac{S_1}{2}$, its value when the screen is removed. According to this there should be a bright spot in the centre of the shadow of a circular disk cast by a distant small source of light. Fresnel found this to be the case.

If the screen has a straight edge and covers half the area $ABCD$ so that the point O is on its edge, then the screen cuts off the light from half of each of the Fresnel zones so that the amplitude at P is $\dfrac{S_1}{4}$ and the illumination is one-quarter of that when the screen is removed. By considering what parts of the zones are cut off it can be shown that the illumination near the boundary of the shadow of a screen with a straight edge should vary in exactly the way which is observed and which was described earlier in this chapter.

One of the best ways of finding experimentally the wave length of light is by means of what is called a diffraction grating. Diffraction gratings are made by ruling a series of equidistant parallel lines either on a plate of glass or on a plane mirror. Gratings ruled on glass are called transmission gratings and we shall consider these only here.

Diffraction Gratings.

In making a transmission grating a plate of plane glass of uniform thickness is taken and the lines are ruled on it with a diamond point which scratches the glass. The lines are ruled by a machine called a dividing engine, which rules the lines straight and very accurately equidistant. The lines are ruled very close together and consist of exceedingly narrow scratches. Gratings have been made with 20,000 or more lines to the centimetre. A useful grating has an area about 3 cms. square ruled with about 15,000 lines or 5000 lines per cm.

In Fig. 66 let PQ be a small part of a grating drawn on a greatly enlarged scale. PQ represents a section of the glass plate and AB, CD, EF represent cross sections of the lines or scratches the lengths of which are supposed to be perpendicular to the plane of the paper. The transparent spaces between the lines are represented by BC, DE, FG, etc. The distance from one side of a space to the same side of the next space, or the sum of the widths of a line and a space, is called the grating space. Let this be denoted by d. If the grating contains n lines and the distance from the first line to the last line is D, then $d = \dfrac{D}{n-1}$. In the figure $d = AC = CE = EG$.

Suppose that a beam of parallel light of wave length λ falls on the grating in the direction from left to right and perpendicular

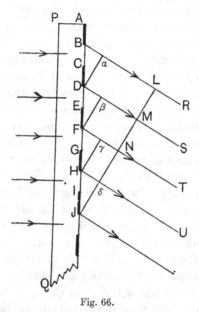

Fig. 66.

to the plane of the grating. The waves in the light are planes parallel to the grating, so that each wave enters all the transparent spaces or slits at the same instant. Draw a series of parallel lines BR, DS, FT, HU, one from the top side of each slit. Drop perpendiculars $D\alpha$, $F\beta$, $H\gamma$ from D, F, H on to BR, DS, FT. A light wave falling on the grating arrives at B, D, F, H at the same instant. The disturbance due to a particular wave travels out from the slits in all directions. Consider the trains of waves which travel along the parallel lines BR, DS, FT, HU. Draw $J\delta NML$ at right angles to BR. At the line JL the train of waves travelling along BR has gone a distance BL from B and the train from D has gone a distance DM. Now $BL - DM = B\alpha$. Also $B\alpha = BD \sin BD\alpha$. But $BD = d$, so that $B\alpha = d \sin \theta$, where θ is the angle between the parallel lines like BR and the normal to the surface of the grating. In the same way the waves at M have travelled a distance equal to $d \sin \theta$ further than the waves at N. Thus if we consider parallel rays from the top of each slit at a plane perpendicular to the rays, the waves from the top of each slit have travelled a distance greater than those from the next lower slit by $d \sin \theta$. In the same way we may consider rays from the middle of each slit or from any corresponding points in each slit. The rays from any slit have to go $d \sin \theta$ further than the corresponding rays from the next slit to reach the plane represented by JL. Suppose now that a convex lens is placed in front of the grating parallel to the plane JL. All the rays from the grating parallel to BR will be brought to a focus on the axis of this lens. At the focus, then, we shall have a series of wave trains which have travelled distances differing by multiples of $d \sin \theta$. If $d \sin \theta$ is equal to λ or a whole number of times λ, say $n\lambda$, then the trains of waves from each slit will agree in phase with those from the corresponding points in all the other slits, so that there will be a bright point at the focus of the lens. If $d \sin \theta$ is not exactly equal to $n\lambda$, then the light from the different slits will interfere and there will be no light at the focus. For example suppose $d \sin \theta = 1 \cdot 001 \lambda$. Then the light from the first slit will be behind that from the 501st slit by a distance equal to $500 \times 1 \cdot 001 \lambda = 500 \cdot 5 \lambda$ so that the light from the slit number 501 will destroy that from the first slit. In the same way the light from the 502nd slit will destroy that from the second slit and so on. Thus unless $d \sin \theta$ is very accurately equal to $n\lambda$ there will be no light at the focus of the lens.

Fig. 67 shows the apparatus used to measure the wave length of the light emitted by a source S with a diffraction grating. The light from S is focused on the slit L of a collimator LO. The collimator is adjusted so that the distance LO is equal to the focal length of the lens at O. The light from each point on the slit L then forms a beam of parallel rays after passing through the colli- mator. The grating GG' is put up with its plane perpendicular to the axis of the collimator and its lines parallel to the slit. The light coming from the grating is examined with a telescope TE, focused for parallel rays. The axis of this telescope lies in a plane perpendicular to the grating lines and containing the axis of the collimator. The telescope can be rotated in this plane about an axis coinciding with the central line on the grating. The angle θ between the axis of the telescope and a normal to the grating can be read on a graduated circle.

If the source S emits light of wave length λ and the angle θ is made equal to one of the roots of the equation $n\lambda = d \sin \theta_n$, an image of the slit L is formed at I on the axis of the telescope. For each point on the slit gives a parallel beam and each of these

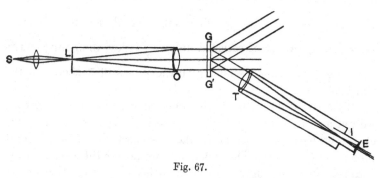

Fig. 67.

beams produces a bright point at the focus of the telescope when $n\lambda = d \sin \theta_n$. The image of the slit is observed through the eyepiece E. The light of wave length λ gives a series of images of the slit. If $n = 0$ we get $\theta_0 = 0$ and there is an image in this position with light of any wave length. This central image is formed by rays which have passed straight through the grating without deviation. If $n = 1$ we get $\lambda = d \sin \theta_1$ so that θ_1 is equal to $\pm \sin^{-1} \dfrac{\lambda}{d}$. There are therefore two images of the slit, one on each side of the central image, making angles of $+ \theta_1$ and $- \theta_1$ with it. If $n = 2$ we get $\theta_2 = \pm \sin^{-1} \dfrac{2\lambda}{d}$ and there are two more images one on each side of the central image and nearly twice as far from it as the first pair.

In the same way there is a pair of images corresponding to each of the other values of n: 3, 4, 5, etc. The wave length of the light can be determined by measuring the values of θ and calculating the value of λ by means of the equation

$$n\lambda = d \sin \theta.$$

For example with the light from a Bunsen flame containing vapour of a thallium salt and a grating for which $d = \frac{1}{5000}$ cm. green images of the slit are observed with the following values of θ:

n	θ	λ in millionths of a mm.
0	0	
1	$\pm 15^\circ$ 31'	535·0
2	$\pm 32^\circ$ 21'	535·1
3	$\pm 53^\circ$ 23'	535·1

If $n \gtreqless 4$ there is no possible value of θ, for the equation $n\lambda = d \sin \theta_n$ cannot be satisfied unless $\dfrac{n\lambda}{d}$ is less than unity. The values of θ in the above table give the values of λ shown in the third column.

If the slit of the collimator is illuminated with white light, then we get a white central image and a series of spectra on each side of it; each of the spectra is like the spectrum of white light produced by a prism. The violet ends are nearer the central image because the wave length of violet light is shorter than that of red light.

A simple way of observing the spectrum of a small source of light is to look at it through a grating held close to the eye without using a collimator or telescope. We then see the source with a series of images of it on each side similar to the images of the slit seen in the telescope.

CHAPTER X

POLARIZATION AND DOUBLE REFRACTION

WHEN a ray of light passes from air into glass, water or many other transparent substances it is refracted according to the law

$$\mu = \frac{\sin i}{\sin r},$$

as we have seen in Chapter II. When a ray enters certain transparent crystals, however, it is found that there are two refracted rays, one of which obeys the ordinary laws of refraction while the other does not. This phenomenon is known as double refraction, and was discovered by Erasmus Bartholinus about the year 1669 with a crystal of Iceland spar.

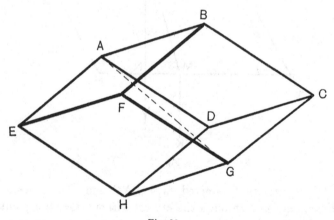

Fig. 68.

Iceland spar is a form of calcium carbonate which is found in Iceland in large transparent crystalline masses. It can be easily split up into rhombohedra, that is into blocks the sides of which are similar parallelograms.

Fig. 68 shows a rhombohedron of Iceland spar. Two diagonally opposite corners at A and G are contained by three equal obtuse angles of 101° 55′ each.   The remaining six corners at B, C, D, E, F and H are contained by one obtuse angle of 101° 55′ and two acute angles of 78° 5′ each.   If the spar has been split so that all the sides of the rhombohedron are of equal length, then the line AG joining the two obtuse corners is equally inclined to the three faces that meet at A and to the three that meet at G.

Suppose a ray of light RP falls on one of the faces of the crystal at P in Fig. 69.   If RP is perpendicular to the surface of the

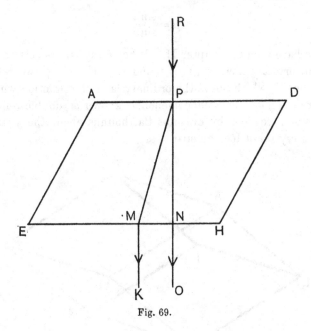

Fig. 69.

crystal we get two refracted rays PNO and PMK.   One, the ordinary ray, goes through the crystal without deviation; but the other PMK is deviated at P and M and emerges parallel to RP. The ray PMK is called the extraordinary ray.   The two rays are of equal intensity.   If the incident ray RP is not perpendicular to the surface of the crystal the ordinary ray is refracted according to the ordinary laws, and the refractive index of the spar for it is about 1·66.   The extraordinary ray is refracted according to quite different laws which need not be considered here.

If the rhombohedron of Iceland spar is laid over a black dot on a sheet of white paper, then on looking at the dot through the spar two images of it are seen. If the spar is turned round, one of these images remains fixed, and the other moves round with the spar.

Huygens discovered that both the ordinary and extraordinary rays after emerging from the crystal possess remarkable properties different from those of ordinary light. The two rays are rather near together, so that it is convenient to get rid of one of them in order that the properties of the other may be examined separately.

A device which enables the ordinary ray to be stopped was invented by Nicol in 1828 and is called Nicol's prism. A crystal of Iceland spar is split so that a rhombohedron at least three times as long as it is wide is obtained, as shown in Fig. 70. This crystal

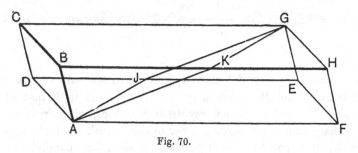

Fig. 70.

is sawn in two at the plane $AJGK$, and the two surfaces are polished and connected together again with Canada balsam. If now a ray of light enters the end $ABCD$ parallel to the length of the prism it is split up into ordinary and extraordinary rays, which fall on the surface of the thin layer of Canada balsam. The refractive index of the balsam is greater than that of the spar for the ordinary ray and this ray is totally reflected from the balsam surface, but the extraordinary ray passes through the balsam and emerges from the prism through the face $EFHG$. The long sides of the prism are blackened so that the ordinary ray is absorbed by them.

Fig. 71 shows the paths of the two rays RPO and RPX through the prism. The ordinary ray RPO is absorbed at O after total reflexion at the plane $GKAJ$.

Fig. 72 shows an arrangement that may be used to examine the properties of the extraordinary ray from a Nicol's prism. S is a source like an arc lamp the rays from which are focused on to a small circular hole in a screen P. The rays passing through this hole pass through the Nicol prism N and then through a convex lens L, which forms an image of the hole in P on a white screen M.

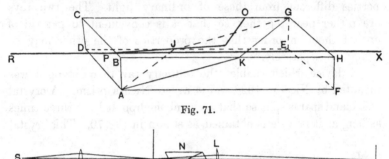

Fig. 71.

Fig. 72.

If a rhombohedron of Iceland spar is put in the path of the rays between L and M we get in general two images at M instead of one. The extraordinary ray from the Nicol gives rise to an ordinary and an extraordinary ray in the crystal. Let the crystal of Iceland spar be turned so that its sides are parallel to the corresponding sides of the Nicol prism and then let it be rotated about an axis parallel to the rays of light between L and M through an angle ϕ.

The following table gives the intensity of illumination or brightness of the images on the screen for different values of the angle ϕ.

ϕ	Ordinary Image	Extraordinary Image
0	0	1
45°	$\frac{1}{2}$	$\frac{1}{2}$
90°	1	0
135°	$\frac{1}{2}$	$\frac{1}{2}$
180°	0	1
225°	$\frac{1}{2}$	$\frac{1}{2}$
270°	1	0
315°	$\frac{1}{2}$	$\frac{1}{2}$
360° or 0°	0	1

The ordinary image remains fixed but the extraordinary image turns round with the crystal. All the light goes into the extraordinary ray when the crystal is parallel to the Nicol, that is, when $\phi = 0$ or $180°$; and all the light goes into the ordinary ray when the crystal is at right angles to the Nicol, that is, when $\phi = 90°$ or $270°$. In intermediate positions both rays appear, and their intensities vary gradually as the crystal is slowly turned round.

This experiment shows that the extraordinary ray has properties which differ for different directions in a plane perpendicular to the ray.

The light in the extraordinary ray is said to be plane polarized. The ordinary ray is also found to possess precisely similar properties in directions at right angles to those of the extraordinary ray. If instead of the Nicol prism and the crystal we use two Nicol prisms in the experiment described above, then of course the ordinary ray produced in the second Nicol is stopped inside it so that only the extraordinary image is formed on the screen. When the two Nicols are parallel, that is, when $\phi = 0$ or $180°$, the image has its maximum brightness, and when they are at right angles or crossed the image disappears. It is clear that the plane polarized rays have properties which are symmetrical with respect to two perpendicular planes containing the rays.

According to the wave theory, light consists of waves travelling along through the ether in a way analogous to the way in which sound waves travel through air. So far, in discussing interference and diffraction, we have not found it necessary to consider whether light waves are longitudinal like sound waves or transverse like waves in a stretched string.

To explain the properties of plane polarized light it is necessary to suppose that light waves are transverse waves, and that in plane polarized light the transverse disturbances in the ether are all parallel to one plane containing the light rays.

Ordinary unpolarized light consists of transverse waves the disturbances in which are not confined to one plane, so that as the series of waves passes a fixed point the direction of the disturbance at the point continually changes and is as often in one direction as in another.

When unpolarized light enters a crystal like Iceland spar the disturbances in the light are resolved parallel to two perpendicular directions and those parallel to one direction are refracted in a different way to those parallel to the other. Thus the unpolarized light is separated into two plane polarized beams polarized in perpendicular planes.

The action of two Nicol prisms on a ray of light may be clearly illustrated by passing a stretched string through two narrow slits in which it can just move freely. Such an arrangement is shown in Fig. 73. If the string at A is shaken about in all directions

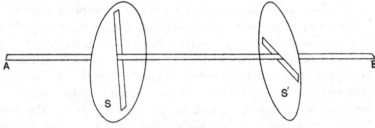

Fig. 73.

transverse to its length, the transverse waves produced will be stopped by the slit S except those in which the displacement is parallel to the slit. The string between S and S' will therefore vibrate parallel to S. If S' is parallel to S the waves between S and S' will get through to B, but if S' is at right angles to S they will all be stopped. In the same way when the two Nicol prisms are parallel, the plane polarized light from the first gets through, and when they are crossed it is stopped entirely.

The fact that the vibrations in light are transverse, that is to say, the direction of the disturbance lies in a plane perpendicular to the ray, was first discovered by Fresnel and Young.

REFERENCE

The Theory of Optics, Drude.

CHAPTER XI

WHEN light falls on the surface of a black body it is absorbed and completely disappears. It is found that the black body gets hotter, so that the light has been converted into heat. When light is emitted by a very hot body like the filament of an incandescent electric lamp, the hot body loses heat. In this case heat is converted into light. We have seen that heat is a form of energy, so that the same is true of light.

A train of sound waves in air possesses energy. When a tuning fork is vibrating it alternately compresses and rarefies the air and sets it in motion. The fork does work on the air and so gives it energy. This energy travels through the air in the sound waves. In the same way at the surface of a very hot body the vibrating atoms produce disturbances in the ether which travel away as light waves. The vibrating atoms do work on the ether and give it energy which takes the form of light.

If the light from a powerful source like an arc lamp is focused by a large convex lens on to a small vessel of water some of the light is absorbed and the water soon begins to boil.

To study the amount of energy in light by measuring the heat produced when it is absorbed by a black body it is convenient to have some means of measuring very small changes of temperature, because the heating effect is small unless the light is very intense.

A convenient instrument for this purpose is the thermopile, of which a modern form is shown in Fig. 74. S_1S_1, S_2S_2, S_3S_3, etc. are bars of silver each about 1 cm. long. B_1B_1, B_2B_2, etc. are bars of bismuth. P_1, P_2, P_3, etc. are small square plates of blackened silver which are each soldered on to the middles of a silver and a bismuth bar.

The bismuth bar at the lower edge of each plate is soldered at its ends to the ends of the silver bar at the upper edge of the next plate. The top silver bar S_1S_1 is connected by copper wires S_1C and S_1C to a wire leading to a galvanometer G and the bottom bismuth bar is connected in a similar way to the other wire leading to the galvanometer. The thermopile is contained in a box in one of the walls of which there is a slit which is opposite to the row of blackened silver plates so that light entering the slit falls on these plates but not on the other parts of the pile. The

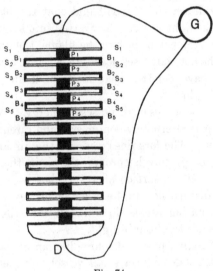

Fig. 74.

light heats the plates, and this causes an electric current to flow round the circuit containing the pile and galvanometer. This current can be measured by the galvanometer. With a sensitive galvanometer a very small amount of energy given to the pile produces a measurable current. The current is approximately proportional to the energy received by the pile per second. If the light from a candle or an electric lamp is allowed to fall on such a thermopile it is found that the current produced varies inversely as the square of the distance between the pile and the lamp. This shows that the current is proportional to the energy received by the pile in unit time, for we know that the light waves from a small source are spheres, so that the energy per unit area in a wave must vary inversely as the square of its radius.

If the thermopile is put up in the spectrum of the light from an arc lamp we can compare the amounts of energy in the different parts of the spectrum.   The apparatus shown in Fig. 32 may be used to produce the spectrum and the thermopile should be put up so that the images of the slit are formed on it.

It is found that the amount of energy is smallest at the violet end of the spectrum and increases steadily as the pile is moved from the violet end to the red end.

If the pile is moved beyond the visible ends of the spectrum it still receives energy, which shows that radiation is present beyond each end of the spectrum although nothing can be seen.   The wave length of the light increases from the violet to the red along the spectrum, so that we should naturally expect the invisible radiation beyond the red to have longer wave lengths than red light, and the invisible radiation beyond the violet to have shorter wave lengths than violet light.   The wave lengths have been measured with diffraction gratings and this expectation has been verified.

The invisible radiation with longer wave lengths than red light is called infra-red light or radiation because it lies beyond the red end of the spectrum, and the invisible radiation with shorter wave lengths than violet light is usually called ultra-violet light because it is beyond the violet end.

It appears that the eye is only affected by light waves the lengths of which lie between certain limits, and that longer or shorter waves do not produce any sensation of sight.   The invisible light waves have all the physical properties of visible light so that they are physically of the same nature as visible light.

In the same way the ear can detect only musical notes with frequencies between certain limits.   Notes of higher and lower frequencies can be obtained and detected, but they do not produce any sensation in the ear.

The distribution of energy in the spectrum of light from a hot black body depends on the temperature of the body.   As the temperature rises the total energy of the light emitted increases rapidly, and the distribution of the energy in the spectrum changes. As the temperature of a body rises it first appears red, then yellowish and finally white or even bluish.

In Fig. 75 the distribution of the energy in the spectrum of the light emitted by a black body at several temperatures is shown graphically. The wave lengths are plotted horizontally and the relative energies vertically.

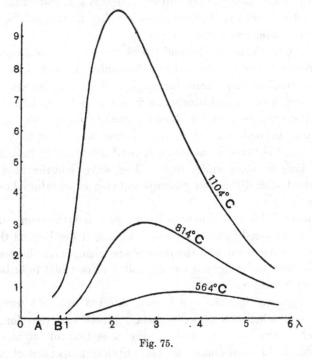

Fig. 75.

The wave lengths are in thousandths of a millimetre and the energy is in arbitrary units. The limits of the visible spectrum are marked A and B on the scale of wave lengths. The upper of the three curves shows the distribution of the energy in the spectrum of the light from a black body at a temperature of 1104°C. At this temperature a body is bright red-hot but not white-hot. The greatest amount of energy in this case is around the wave length 0·0021 mm. At 814° C. the maximum is near 0·0026 mm. and at 564° C. near 0·0035 mm. The maximum energy occurs at a wave length 0·0006 mm. which is that of yellow light when the temperature is about 4500° C.

The sources of artificial white light which are now available all have temperatures very much below 4500° C., so that they emit far more energy in the form of invisible infra-red light than in the form of visible light. There is much more energy beyond the red in the spectrum of the light from an arc lamp than there is in the visible part of the spectrum.

The energy of the ultra-violet light from an arc light is comparatively small. It can only be detected with a very sensitive thermopile and galvanometer. There are however other methods of detecting ultra-violet light.

It is found that certain substances when exposed to ultra-violet light emit visible light. This phenomenon is called fluorescence. For example barium platinocyanide, which is a pale yellow substance, when exposed to ultra-violet light becomes luminous emitting green light. If a piece of paper coated with a thin layer of barium platinocyanide is put up beyond the violet end of the spectrum of an arc lamp it emits green light, thus showing the presence of the ultra-violet light. Ultra-violet light is strongly absorbed by flint glass, so that if a flint glass lens or prism is used to produce the spectrum very little ultra-violet light will be found in it. Quartz is more transparent to ultra-violet light than glass, so that to obtain a spectrum rich in ultra-violet light it is best to use lenses and prisms made of quartz.

REFERENCE

Physical Optics, R. W. Wood.

INDEX

Printed in the United States
By Bookmasters